THE GROSSET WORLD ATLAS

GROSSET & DUNLAP
A NATIONAL GENERAL COMPANY
NEW YORK

GAZETTEER-INDEX OF THE WORLD

This alphabetical list of grand divisions, countries, states, colonial possessions, etc., gives area, population, capital or chief town, and index references and page numbers on which they are shown on the largest scale. The index reference shows the square on the respective map in which the name of the entry may be located.

Country	Area (Sq. Miles)	Population	Capital or Chief Town	Index Ref.	Plate No.
Afars and Issas, Terr.	8,498	125,050	Djibouti	P 9	46
*Afghanistan	250,000	17,078,263	Kabul	B 2	34
Africa	11,682,000	345,000,000	......		46,47
Alabama, U.S.A.	51,609	3,444,165	Montgomery	M 6	7
Alaska, U.S.A.	586,412	302,173	Juneau	F 8	6
*Albania	11,100	2,126,000	Tiranë	D 5	26
Alberta, Canada	255,285	1,614,000	Edmonton	E 5	3
*Algeria	919,515	13,547,000	Algiers	G 4	46
American Samoa	76	27,769	Pago Pago	J 7	44
Andorra	175	19,000	Andorra la Vella	G 1	25
Angola	481,351	5,430,000	Luanda	J13	47
Antarctica	5,500,000	....			48
Antigua	171	63,000	St. Johns	G 3	12
*Argentina	1,072,070	23,983,000	Buenos Aires	H10	17
Arizona, U.S.A.	113,909	1,772,482	Phoenix	E 6	6
Arkansas, U.S.A.	53,104	1,923,295	Little Rock	K 6	7
Ascension	34	1,486	Georgetown	D13	47
Asia	17,032,000	2,043,997,000	......		31
*Australia	2,967,741	12,630,000	Canberra		42
*Austria	32,374	7,419,341	Vienna	K 6	18
Bahama Islands	4,404	168,838	Nassau	C 1	12
*Bahrain	231	207,000	Manama	E 4	32
Bangladesh	55,126	70,000,000	Dacca	F 4	34
*Barbados	166	253,620	Bridgetown	G 4	12
*Belgium	11,779	9,660,154	Brussels	H 5	18
Bermuda	21	52,000	Hamilton	G 3	12
*Bhutan	18,000	770,000	Thimphu	G 3	34
*Bolivia	24,163	4,804,000	La Paz, Sucre	G 7	16
Botswana	219,815	629,000	Gaborone	M16	47
*Brazil	3,284,426	90,840,000	Brasília	L 7	16
British Columbia, Canada	366,255	2,161,000	Victoria	D 6	3
British Honduras	8,867	122,000	Belmopan	C 2	10
British Indian Ocean Terr.	30	1,000	Victoria (Seychelles)	L10	31
Brunei	2,226	130,000	Bandar Seri Begawan	E 4	38
*Bulgaria	42,829	8,501,000	Sofia	F 4	26
*Burma	261,789	27,000,000	Rangoon	C 3	40
*Burundi	10,747	3,475,000	Bujumbura	M12	47
California, U.S.A.	158,693	19,953,134	Sacramento	C 5	6
*Cambodia	69,698	6,701,000	Phnom Penh	E 5	40
*Cameroon	183,568	5,836,000	Yaoundé	J10	46
*Canada	3,851,809	21,489,000	Ottawa		3
Canal Zone	647	44,650	Balboa Heights	G 6	10
Cape Verde Islands	1,557	250,000	Praia	H 5	1
Cayman Islands	100	10,652	Georgetown	B 3	12
*Central African Republic	240,534	1,518,000	Bangui	K11	46
Central America	196,928	16,090,000	....		10
*Ceylon (Sri Lanka)	25,332	12,300,000	Colombo	D 7	34
*Chad	495,753	3,510,000	Fort-Lamy	K 9	46
Channel Islands	75	117,000	St. Helier	E 6	20
*Chile	292,257	8,834,820	Santiago	F10	17
*China (mainland)	3,691,506	740,000,000	Peking		36
China (Taiwan)	13,948	14,577,000	Taipei	K 7	36
*Colombia	439,513	21,117,000	Bogotá	F 3	16
Colorado, U.S.A.	102,247	2,207,259	Denver	G 5	6
Comoro Is.	838	270,000	Moroni	P14	47
*Congo, Rep. of	132,046	915,000	Brazzaville	J12	47
Connecticut, U.S.A.	5,009	3,032,217	Hartford	P 4	7
Cook Islands	93	20,000	Avarua	L 8	44
*Costa Rica	19,575	1,800,000	San José	F 5	10
*Cuba	44,206	8,553,395	Havana	A 2	12
*Cyprus	3,473	649,000	Nicosia	B 2	32
*Czechoslovakia	49,370	14,497,000	Prague	K 6	18
*Dahomey	44,290	2,640,000	Porto-Novo	G10	46
Delaware, U.S.A.	2,057	548,104	Dover	P 5	7
*Denmark	16,625	4,910,000	Copenhagen	G 9	23
District of Columbia, U.S.A. ..	67	756,510	Washington	O 5	7
Dominica	290	70,302	Roseau	G 4	12
*Dominican Republic	18,704	4,011,589	Santo Domingo	E 3	12
*Ecuador	109,483	6,144,000	Quito	E 4	16
*Egypt	386,100	32,501,000	Cairo	N 5	46
*El Salvador	8,260	3,418,455	San Salvador	C 4	10

Country	Area (Sq. Miles)	Population	Capital or Chief Town	Index Ref.	Plate No.
England, U.K.	50,327	46,102,300	London		20
*Equatorial Guinea	10,832	286,000	Santa Isabel	G11	46
*Ethiopia	471,776	24,764,000	Addis Ababa	O10	46
Europe	4,063,000	644,574,000	......		18-30
Faerøe Islands, Den.	540	38,000	Tórshavn	F 3	18
Falkland Islands	4,618	2,000	Stanley	J14	17
*Fiji	7,015	519,000	Suva	H 8	44
*Finland	130,128	4,706,000	Helsinki		23
Florida, U.S.A.	58,560	6,789,443	Tallahassee	M 7	7
*France	212,841	50,770,000	Paris		21
French Guiana	35,135	48,000	Cayenne	K 2	16
French Polynesia	1,544	109,000	Papeete	M 7	44
*Gabon	103,346	500,000	Libreville	H11	47
*Gambia	4,003	357,000	Bathurst	C 9	46
Georgia, U.S.A.	58,876	4,589,575	Atlanta	M 6	7
Germany, East (German Democratic Republic)	41,814	17,117,000	Berlin		22
Germany, West (Federal Republic of)	95,959	61,194,600	Bonn		22
*Ghana	91,843	8,545,561	Accra	G10	46
Gibraltar	2	27,000	Gibraltar	D 4	25
Gilbert and Ellice Is.	369	55,185	Bairiki	H 5	44
*Great Britain and Northern Ireland (United Kingdom)..	94,214	55,534,000	London		20
*Greece	50,548	8,838,000	Athens	F 7	26
Greenland	840,000	47,000	Godthåb	N 3	2
Grenada	133	105,000	St. George's	F 4	12
Guadeloupe and Dependencies	687	324,000	Basse-Terre	F 4	12
Guam	209	86,926	Agaña		44
*Guatemala	42,042	5,200,000	Guatemala	B 3	10
*Guinea	94,925	3,890,000	Conakry	D10	46
*Guyana	83,000	763,000	Georgetown	J 2	16
*Haiti	10,694	4,867,190	Port-au-Prince	D 3	12
Hawaii, U.S.A.	6,450	769,913	Honolulu	H 8	6
*Holland (Netherlands)	13,958	13,077,000	Amsterdam, The Hague	H 5	18
*Honduras	43,277	2,495,000	Tegucigalpa	D 3	10
Hong Kong	398	4,089,000	Victoria	H 7	36
*Hungary	35,915	10,315,597	Budapest	L 6	18
*Iceland	39,768	203,000	Reykjavík	H 2	1
Idaho, U.S.A.	83,557	713,008	Boise	E 4	6
Illinois, U.S.A.	56,400	11,113,976	Springfield	L 5	7
*India	1,261,483	546,955,945	New Delhi		34
Indiana, U.S.A.	36,291	5,193,669	Indianapolis	M 5	7
*Indonesia	763,264	119,572,000	Djakarta		38
Iowa, U.S.A.	56,290	2,825,041	Des Moines	K 4	7
*Iran	636,293	28,448,000	Tehran	F 2	32
*Iraq	167,924	9,431,000	Baghdad	D 3	32
*Ireland	26,600	2,944,000	Dublin	C 4	20
Ireland, Northern, U.K.	5,459	1,512,500	Belfast	C 3	20
Isle of Man, U.K.	227	50,000	Douglas	D 3	20
*Israel	7,993	2,911,000	Jerusalem	C 3	32
*Italy	116,303	54,504,000	Rome		24
*Ivory Coast	124,503	4,800,000	Abidjan	E10	46
*Jamaica	4,411	1,972,000	Kingston	C 3	12
*Japan	143,662	103,529,000	Tokyo	N 4	36
*Jordan	37,297	2,300,000	Amman	C 3	32
Kansas, U.S.A.	82,264	2,249,071	Topeka	K 5	7
Kentucky, U.S.A.	40,395	3,219,311	Frankfort	M 5	7
*Kenya	224,960	10,880,200	Nairobi	O12	47
Korea, North	46,540	13,300,000	P'yŏngyang	L 4	36
Korea, South	38,452	31,683,000	Seoul	L 4	36
*Kuwait	6,177	733,196	Al Kuwait	E 4	32
*Laos	91,459	2,900,000	Vientiane	D 2	40
*Lebanon	4,015	2,800,000	Beirut	B 3	32
*Lesotho	11,716	930,000	Maseru	M17	47
*Liberia	43,000	1,200,000	Monrovia	D10	46
*Libya	679,359	1,900,000	Tripoli	K 6	46
Liechtenstein	61	21,000	Vaduz	J 6	18
Louisiana, U.S.A.	48,523	3,643,180	Baton Rouge	L 7	7
*Luxembourg	999	339,000	Luxembourg	J 6	18

*Members of the United Nations

(continued following page 48)

Library of Congress Cataloging in Publication Data

Hammond Incorporated.
  Hammond headline world atlas.
  1. Atlases.   I. Title.   II. Title: Headline world atlas.
G1019.H3125   912   72-14224
ISBN 0-8437-2505-2

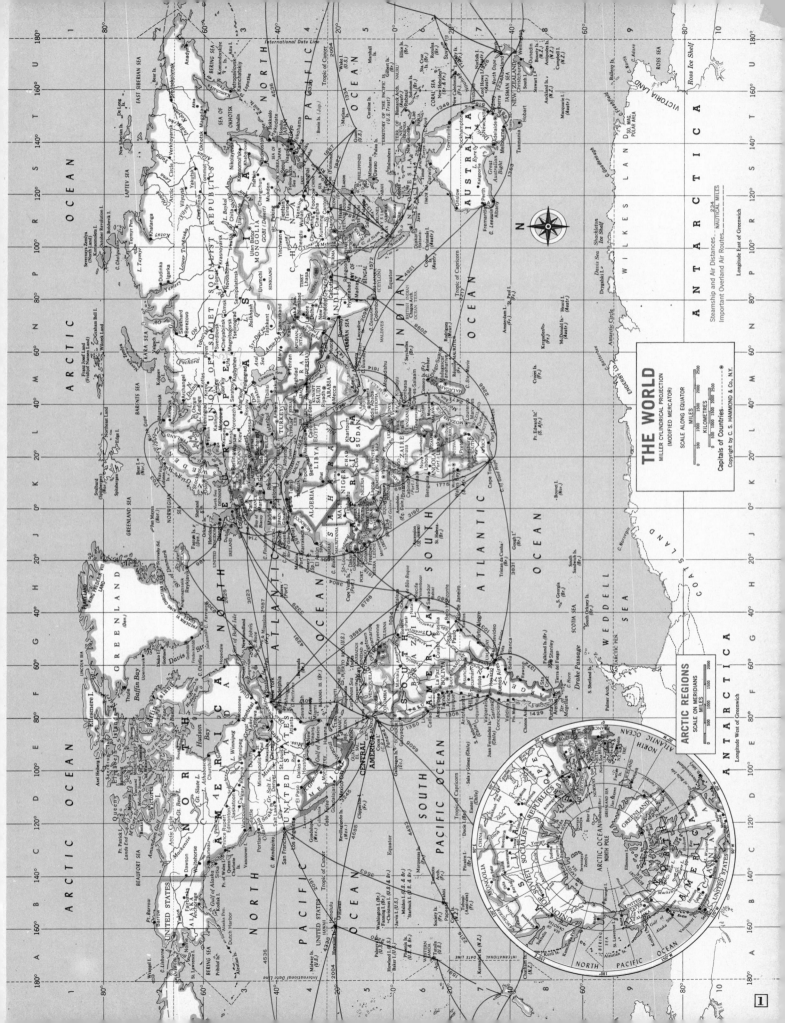

NORTH AMERICA

LAMBERT AZIMUTHAL EQUAL-AREA PROJECTION

SCALE OF MILES
0 100 200 400 600 800

SCALE OF KILOMETRES
0 200 400 600 800

Capitals of Countries..............☆
International Boundaries......._ . _ . _
Other Boundaries................._____
Canals............................

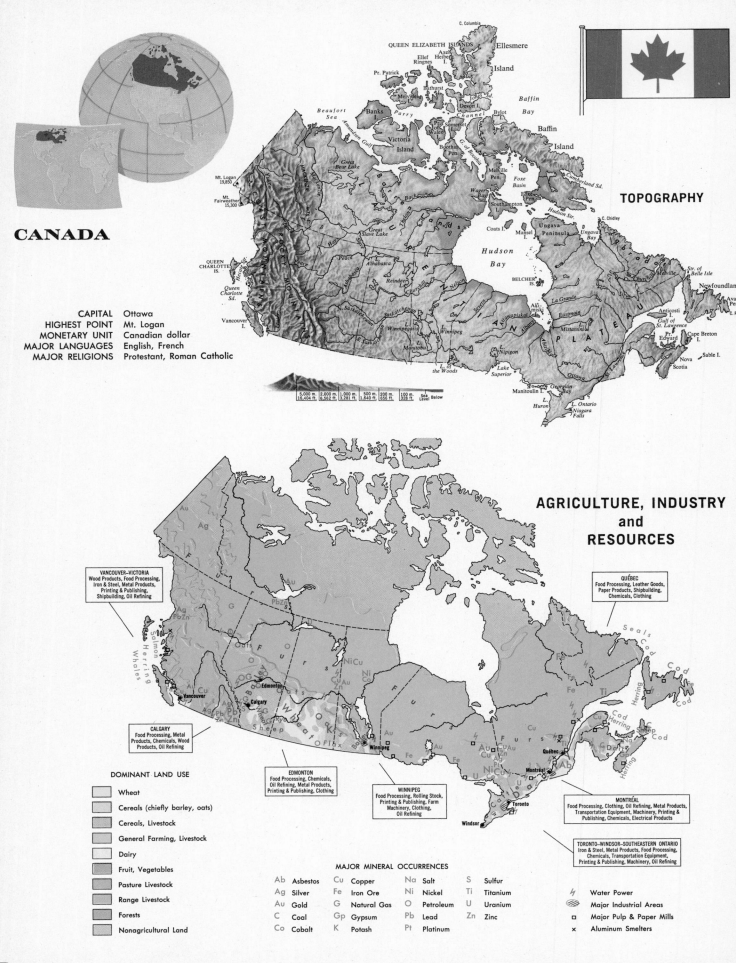

CANADA

QUEEN ELIZABETH ISLANDS

TOPOGRAPHY

CAPITAL — Ottawa
HIGHEST POINT — Mt. Logan
MONETARY UNIT — Canadian dollar
MAJOR LANGUAGES — English, French
MAJOR RELIGIONS — Protestant, Roman Catholic

Mt. Logan 19,850
Mt. Fairweather 15,300

| 5,000 m. 16,404 ft. | 2,000 m. 6,562 ft. | 1,000 m. 3,281 ft. | 500 m. 1,640 ft. | 200 m. 656 ft. | 100 m. 328 ft. | Sea Level | Below |

AGRICULTURE, INDUSTRY and RESOURCES

VANCOUVER–VICTORIA
Wood Products, Food Processing, Iron & Steel, Metal Products, Printing & Publishing, Shipbuilding, Oil Refining

QUÉBEC
Food Processing, Leather Goods, Paper Products, Shipbuilding, Chemicals, Clothing

CALGARY
Food Processing, Metal Products, Chemicals, Wood Products, Oil Refining

EDMONTON
Food Processing, Chemicals, Oil Refining, Metal Products, Printing & Publishing, Clothing

WINNIPEG
Food Processing, Rolling Stock, Printing & Publishing, Farm Machinery, Clothing, Oil Refining

MONTRÉAL
Food Processing, Clothing, Oil Refining, Metal Products, Transportation Equipment, Machinery, Printing & Publishing, Chemicals, Electrical Products

TORONTO–WINDSOR–SOUTHEASTERN ONTARIO
Iron & Steel, Metal Products, Food Processing, Chemicals, Transportation Equipment, Printing & Publishing, Machinery, Oil Refining

DOMINANT LAND USE

- Wheat
- Cereals (chiefly barley, oats)
- Cereals, Livestock
- General Farming, Livestock
- Dairy
- Fruit, Vegetables
- Pasture Livestock
- Range Livestock
- Forests
- Nonagricultural Land

MAJOR MINERAL OCCURRENCES

Ab	Asbestos	Cu	Copper	Na	Salt	S	Sulfur
Ag	Silver	Fe	Iron Ore	Ni	Nickel	Ti	Titanium
Au	Gold	G	Natural Gas	O	Petroleum	U	Uranium
C	Coal	Gp	Gypsum	Pb	Lead	Zn	Zinc
Co	Cobalt	K	Potash	Pt	Platinum		

- ⚡ Water Power
- Major Industrial Areas
- ▫ Major Pulp & Paper Mills
- × Aluminum Smelters

4

TOPOGRAPHY

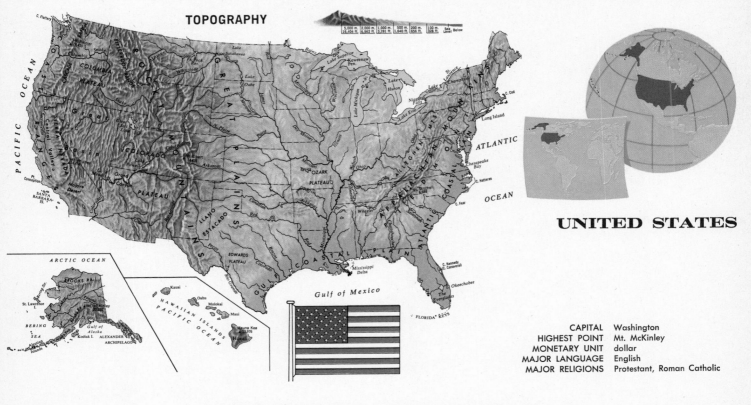

UNITED STATES

CAPITAL	Washington
HIGHEST POINT	Mt. McKinley
MONETARY UNIT	dollar
MAJOR LANGUAGE	English
MAJOR RELIGIONS	Protestant, Roman Catholic

AGRICULTURE, INDUSTRY and RESOURCES

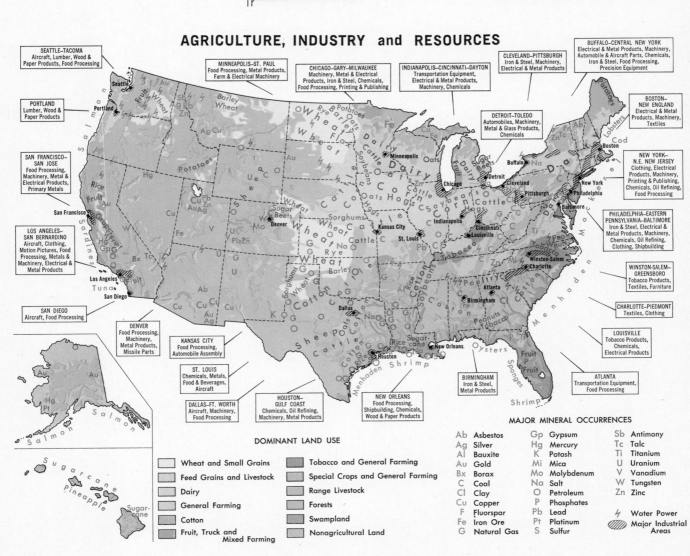

SEATTLE-TACOMA
Aircraft, Lumber, Wood &
Paper Products, Food Processing

PORTLAND
Lumber, Wood &
Paper Products

**SAN FRANCISCO-
SAN JOSE**
Food Processing,
Machinery, Metal &
Electrical Products,
Primary Metals

**LOS ANGELES-
SAN BERNARDINO**
Aircraft, Clothing,
Motion Pictures, Food
Processing, Metals &
Machinery, Electrical &
Metal Products

SAN DIEGO
Aircraft, Food Processing

DENVER
Food Processing,
Machinery,
Metal Products,
Missile Parts

KANSAS CITY
Food Processing,
Automobile Assembly

ST. LOUIS
Chemicals, Metals,
Food & Beverages,
Aircraft

DALLAS-FT. WORTH
Aircraft, Machinery,
Food Processing

**HOUSTON-
GULF COAST**
Chemicals, Oil Refining,
Machinery, Metal Products

NEW ORLEANS
Food Processing,
Shipbuilding, Chemicals,
Wood & Paper Products

BIRMINGHAM
Iron & Steel,
Metal Products

MINNEAPOLIS-ST. PAUL
Food Processing, Metal Products,
Farm & Electrical Machinery

CHICAGO-GARY-MILWAUKEE
Machinery, Metal & Electrical
Products, Iron & Steel, Chemicals,
Food Processing, Printing & Publishing

INDIANAPOLIS-CINCINNATI-DAYTON
Transportation Equipment,
Electrical & Metal Products,
Machinery, Chemicals

DETROIT-TOLEDO
Automobiles, Machinery,
Metal & Glass Products,
Chemicals

CLEVELAND-PITTSBURGH
Iron & Steel, Machinery,
Electrical & Metal Products

BUFFALO-CENTRAL NEW YORK
Electrical & Metal Products, Machinery,
Automobile & Aircraft Parts, Chemicals,
Iron & Steel, Food Processing,
Precision Equipment

**BOSTON-
NEW ENGLAND**
Electrical & Metal
Products, Machinery,
Textiles

**NEW YORK-
N.E. NEW JERSEY**
Clothing, Electrical
Products, Machinery,
Printing & Publishing,
Chemicals, Oil Refining,
Food Processing

**PHILADELPHIA-EASTERN
PENNSYLVANIA-BALTIMORE**
Iron & Steel, Electrical &
Metal Products, Machinery,
Chemicals, Oil Refining,
Clothing, Shipbuilding

**WINSTON-SALEM-
GREENSBORO**
Tobacco Products,
Textiles, Furniture

CHARLOTTE-PIEDMONT
Textiles, Clothing

LOUISVILLE
Tobacco Products,
Chemicals,
Electrical Products

ATLANTA
Transportation Equipment,
Food Processing

DOMINANT LAND USE

- Wheat and Small Grains
- Feed Grains and Livestock
- Dairy
- General Farming
- Cotton
- Fruit, Truck and Mixed Farming
- Tobacco and General Farming
- Special Crops and General Farming
- Range Livestock
- Forests
- Swampland
- Nonagricultural Land

MAJOR MINERAL OCCURRENCES

Ab	Asbestos	Gp	Gypsum	Sb	Antimony
Ag	Silver	Hg	Mercury	Tc	Talc
Al	Bauxite	K	Potash	Ti	Titanium
Au	Gold	Mi	Mica	U	Uranium
Bx	Borax	Mo	Molybdenum	V	Vanadium
C	Coal	Na	Salt	W	Tungsten
Cl	Clay	O	Petroleum	Zn	Zinc
Cu	Copper	P	Phosphates		
F	Fluorspar	Pb	Lead	⚡	Water Power
Fe	Iron Ore	Pt	Platinum	▨	Major Industrial
G	Natural Gas	S	Sulfur		Areas

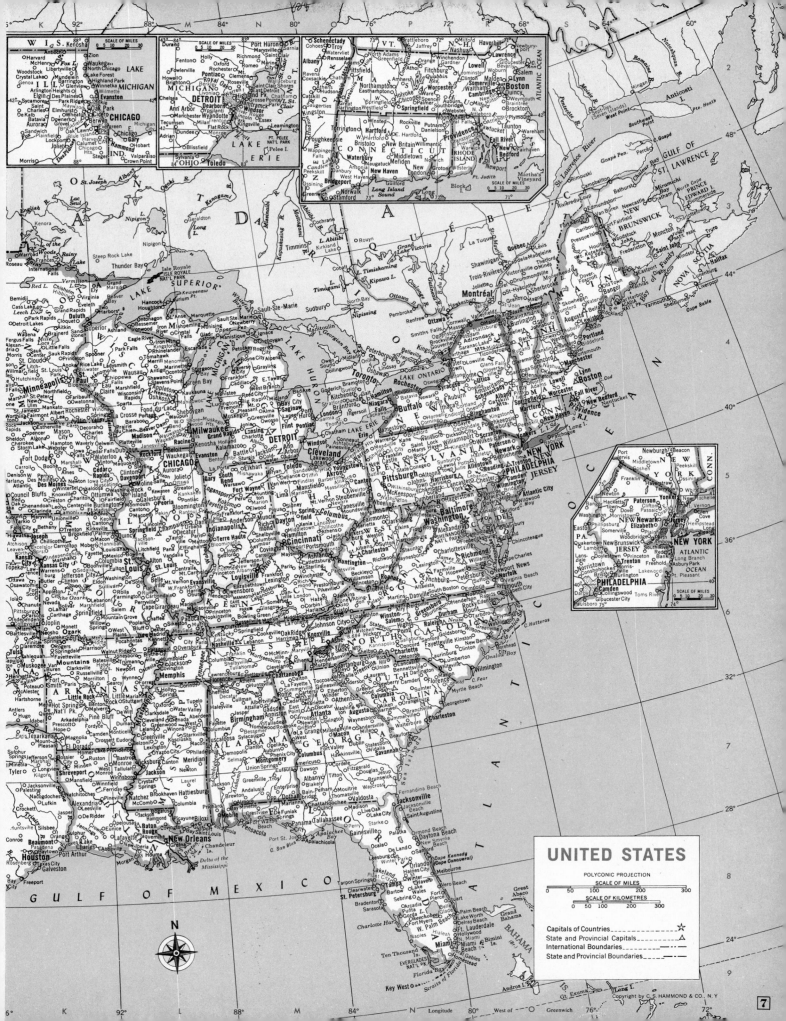

UNITED STATES

POLYCONIC PROJECTION

SCALE OF MILES

0   50   100   200   300

SCALE OF KILOMETRES

0   50   100   200   300

Capitals of Countries	☆
State and Provincial Capitals	△
International Boundaries	
State and Provincial Boundaries	

Copyright by C. S. HAMMOND & CO., N.Y.

7

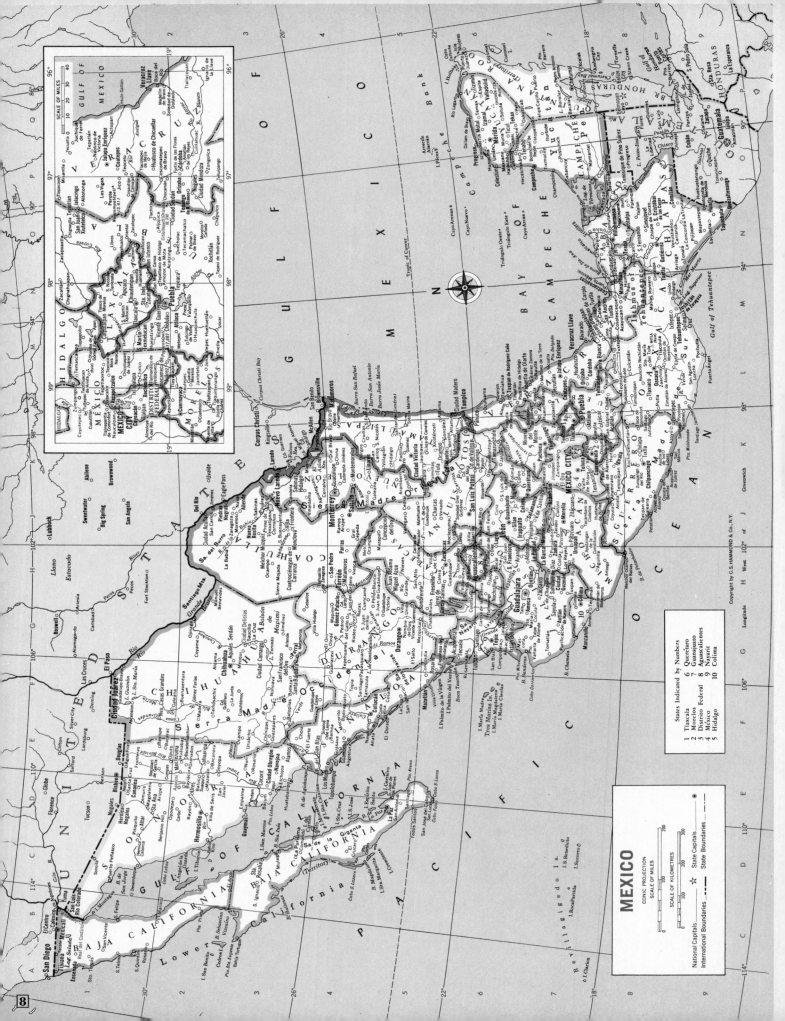

MEXICO

CONIC PROJECTION

SCALE OF MILES

SCALE OF KILOMETRES

National Capitals ....●
State Capitals ....☆
International Boundaries .... – – –
State Boundaries .... – · – · –

States Indicated by Numbers

1 Tlaxcala	6 Querétaro
2 Morelos	7 Guanajuato
3 Distrito Federal	8 Aguascalientes
4 México	9 Nayarit
5 Hidalgo	10 Colima

SCALE OF MILES

GULF OF MEXICO

MEXICO

CAPITAL	Mexico City
HIGHEST POINT	Citlaltépetl
MONETARY UNIT	Mexican peso
MAJOR LANGUAGE	Spanish
MAJOR RELIGION	Roman Catholic

TOPOGRAPHY

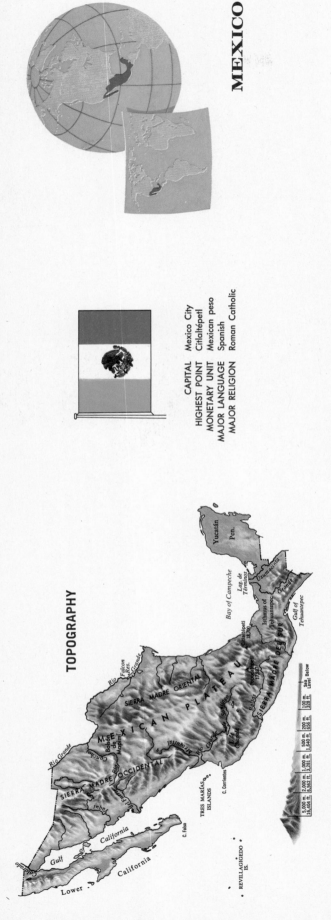

5,000 m. 16,404 ft.
2,000 m. 6,562 ft.
1,000 m. 3,281 ft.
500 m. 1,640 ft.
200 m. 656 ft.
100 m. 328 ft.
Sea Level
Below

AGRICULTURE, INDUSTRY and RESOURCES

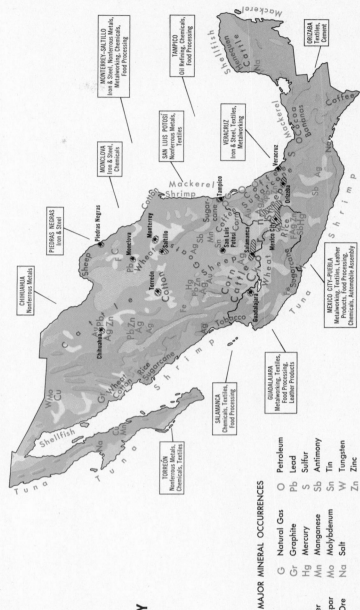

DOMINANT LAND USE

Wheat, Livestock
Cereals (chiefly corn), Livestock
Diversified Tropical Cash Crops
Cotton, Mixed Cereals
Livestock, Limited Agriculture
Range Livestock
Forests
Nonagricultural Land

Water Power
Major Industrial Areas

MAJOR MINERAL OCCURRENCES

Ag	Silver		G	Natural Gas
Au	Gold		Gr	Graphite
C	Coal		Hg	Mercury
Cu	Copper		Mn	Manganese
F	Fluorspar		Mo	Molybdenum
Fe	Iron Ore		Na	Salt

O	Petroleum
Pb	Lead
S	Sulfur
Sb	Antimony
Sn	Tin
W	Tungsten
Zn	Zinc

CHIHUAHUA
Nonferrous Metals

PIEDRAS NEGRAS
Iron & Steel

MONCLOVA
Iron & Steel, Chemicals

MONTERREY–SALTILLO
Iron & Steel, Nonferrous Metals, Metalworking, Chemicals, Food Processing

SAN LUIS POTOSÍ
Nonferrous Metals, Textiles

TAMPICO
Oil Refining, Chemicals, Food Processing

VERACRUZ
Iron & Steel, Metalworking

ORIZABA
Textiles, Cement

TORREÓN
Nonferrous Metals, Chemicals, Textiles

SALAMANCA
Chemicals, Textiles, Food Processing

GUADALAJARA
Metalworking, Textiles, Food Processing, Leather Products

MEXICO CITY–PUEBLA
Metalworking, Textiles, Leather Products, Food Processing, Chemicals, Automobile Assembly

9

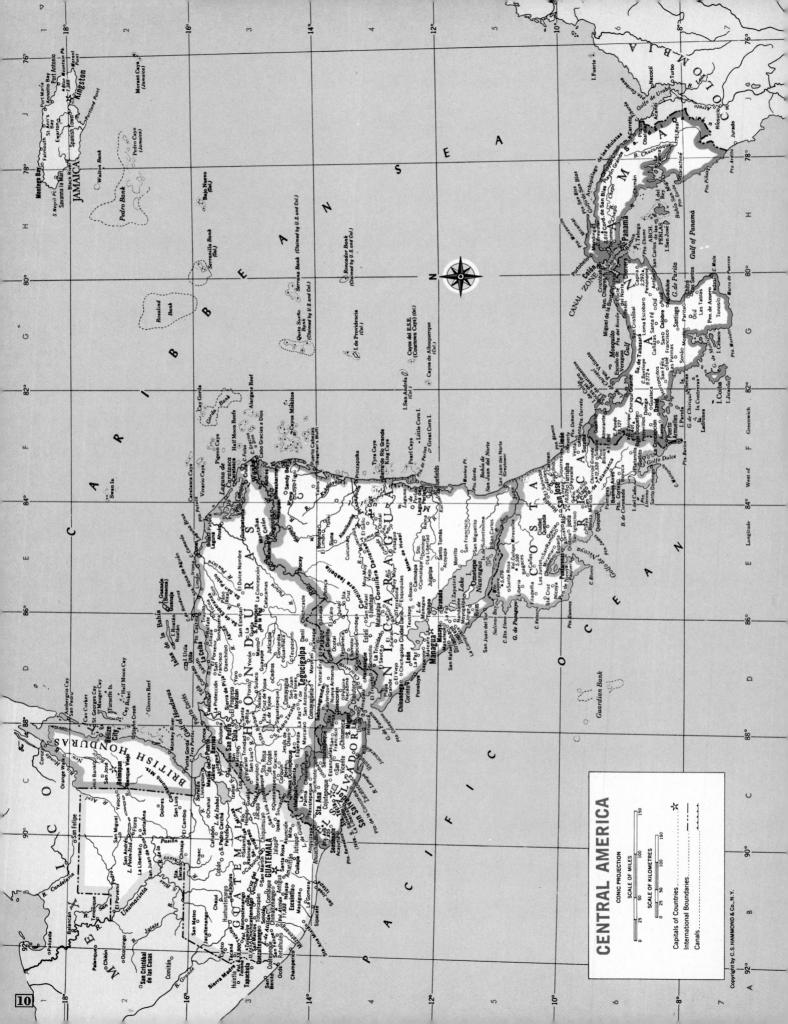

CENTRAL AMERICA

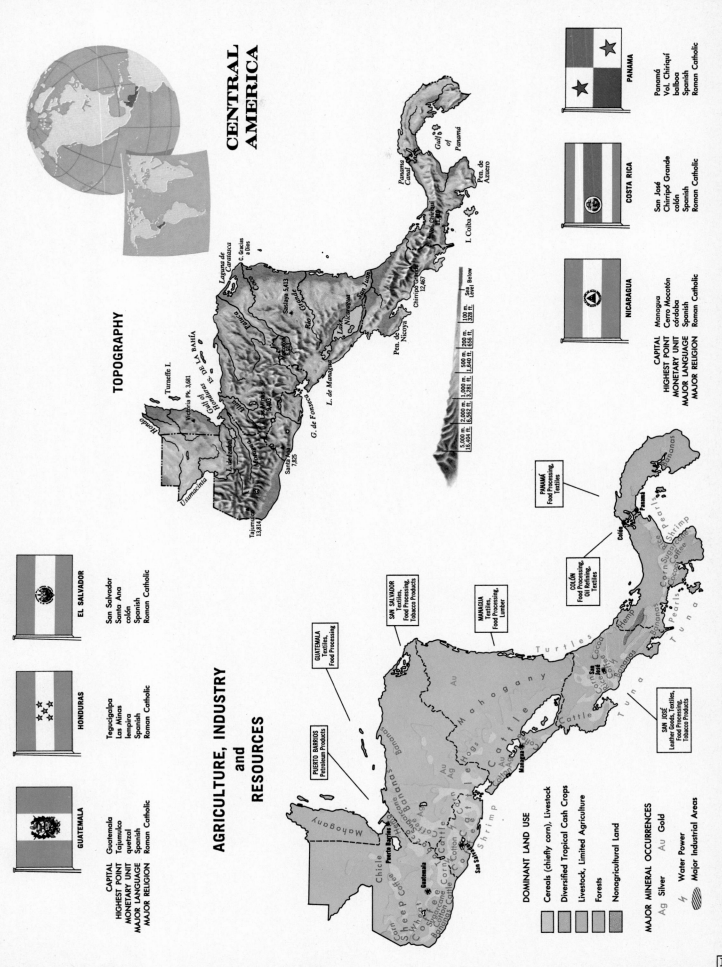

CENTRAL AMERICA

TOPOGRAPHY

Turneffe I.
Victoria Pk. 3,681
Gulf of Honduras
IS. DE LA BAHÍA
Hondo
Las Minas 9,403
Motagua
Izabal
Usumacinta
Santa Ana 7,825
Tajumulco 13,814
G. de Fonseca
L. de Managua
Lake Nicaragua
Pen. de Nicoya
Ulúa
Coco
Río Grande
Saslaya 5,413
San Juan
Patuca
Laguna de Caratasca
C. Gracias a Dios
Chirripó Grande 12,467
Vol. Chiriquí 11,410
I. Coiba
Pen. de Azuero
Panama Canal
Gulf of Panamá

5,000 m. 16,404 ft.	2,000 m. 6,562 ft.	1,000 m. 3,281 ft.
500 m. 1,640 ft.	200 m. 656 ft.	100 m. 328 ft.
Sea Level	Below	

AGRICULTURE, INDUSTRY and RESOURCES

PUERTO BARRIOS
Petroleum Products

GUATEMALA
Textiles,
Food Processing

SAN SALVADOR
Textiles,
Food Processing,
Tobacco Products

MANAGUA
Textiles,
Food Processing,
Lumber

SAN JOSÉ
Leather Goods, Textiles,
Food Processing,
Tobacco Products

COLÓN
Food Processing,
Oil Refining,
Textiles

PANAMÁ
Food Processing,
Textiles

DOMINANT LAND USE

- Cereals (chiefly corn), Livestock
- Diversified Tropical Cash Crops
- Livestock, Limited Agriculture
- Forests
- Nonagricultural Land

MAJOR MINERAL OCCURRENCES

Ag Silver   Au Gold

⋏ Water Power   Major Industrial Areas

GUATEMALA

CAPITAL	Guatemala
HIGHEST POINT	Tajumulco
MONETARY UNIT	quetzal
MAJOR LANGUAGE	Spanish
MAJOR RELIGION	Roman Catholic

HONDURAS

Tegucigalpa
Las Minas
lempira
Spanish
Roman Catholic

EL SALVADOR

San Salvador
Santa Ana
colón
Spanish
Roman Catholic

NICARAGUA

CAPITAL	Managua
HIGHEST POINT	Cerro Mocotón
MONETARY UNIT	córdoba
MAJOR LANGUAGE	Spanish
MAJOR RELIGION	Roman Catholic

COSTA RICA

San José
Chirripó Grande
colón
Spanish
Roman Catholic

PANAMA

Panamá
Vol. Chiriquí
balboa
Spanish
Roman Catholic

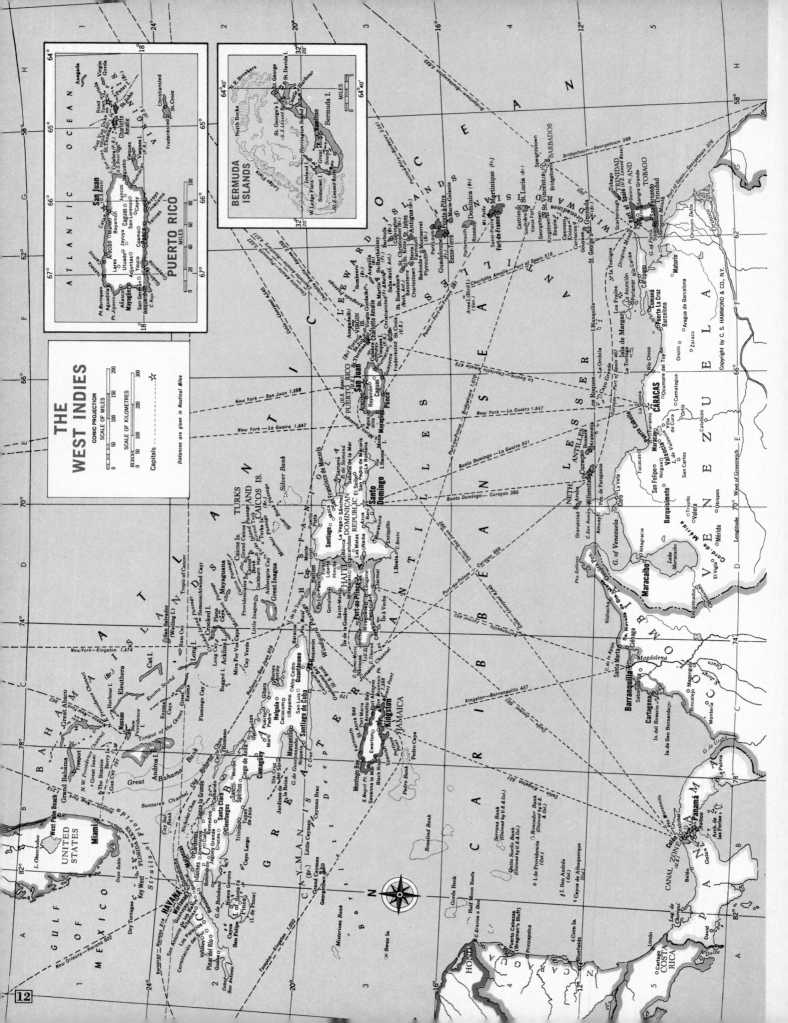

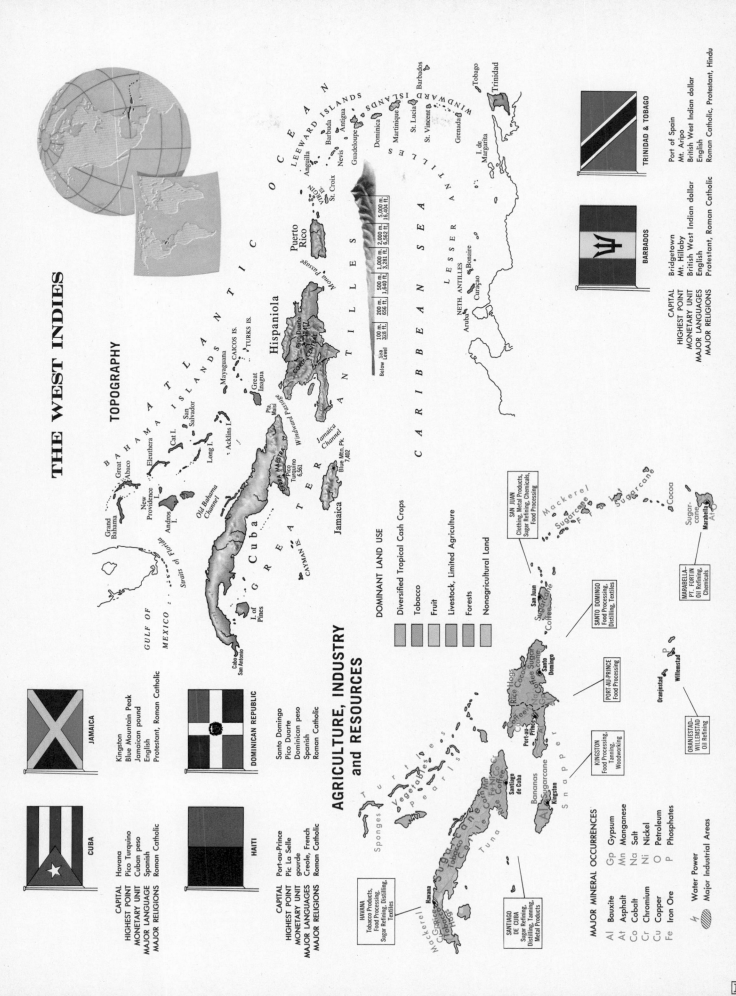

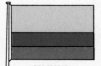

COLOMBIA

VENEZUELA

ECUADOR

PERU

BRAZIL

CAPITAL	Bogotá	Caracas	Quito	Lima	Brasília
HIGHEST POINT	Pico Cristóbal Colón	Pico Bolívar	Chimborazo	Huascarán	Pico da Banderia
MONETARY UNIT	Colombian peso	bolívar	sucre	sol	cruzeiro
MAJOR LANGUAGES	Spanish	Spanish	Spanish, Indian	Spanish, Indian	Portuguese
MAJOR RELIGIONS	Roman Catholic	Roman Catholic	Roman Catholic	Roman Catholic	Roman Catholic

TOPOGRAPHY

5,000 m.	2,000 m.	1,000 m.	500 m.	200 m.	100 m.	Sea	Below
16,404 ft.	6,562 ft.	3,281 ft.	1,640 ft.	656 ft.	328 ft.	Level	

BOLIVIA

PARAGUAY

CAPITAL	La Paz, Sucre	Asunción
HIGHEST POINT	Nevada Ancohuma	Amambay Range
MONETARY UNIT	Bolivian peso	guaraní
MAJOR LANGUAGES	Spanish, Indian	Spanish, Indian
MAJOR RELIGION	Roman Catholic	Roman Catholic

CHILE

ARGENTINA

URUGUAY

CAPITAL	Santiago	Buenos Aires	Montevideo
HIGHEST POINT	Ojos del Salado	Cerro Aconcagua	Mirador Nacional
MONETARY UNIT	Chilean escudo	Argentine peso	Uruguayan peso
MAJOR LANGUAGE	Spanish	Spanish	Spanish
MAJOR RELIGION	Roman Catholic	Roman Catholic	Roman Catholic

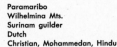

GUYANA

Georgetown
Mt. Roraima
Guyana dollar
English, East Indian
Christian, Hindu, Mohammedan

SURINAM

Paramaribo
Wilhelmina Mts.
Surinam guilder
Dutch
Christian, Mohammedan, Hindu

FRENCH GUIANA

Cayenne
Chaîne de Milthaide
French franc
French
Roman Catholic, Protestant

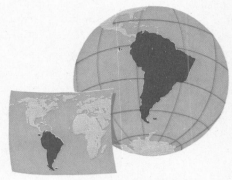

SOUTH AMERICA

AGRICULTURE, INDUSTRY and RESOURCES

AMUAY–PUNTA CARDÓN
Oil Refining

MEDELLÍN
Textiles, Clothing,
Leather Goods

BOGOTÁ
Textiles, Leather Goods,
Cement, Electrical Equipment

CARACAS
Textiles, Chemicals,
Automobiles

CIUDAD
GUAYANA
Iron & Steel,
Aluminum

BELO HORIZONTE
Iron & Steel, Textiles,
Cement, Metal Products

RIO DE JANEIRO
Iron & Steel, Chemicals,
Food Processing, Textiles,
Glass Products,
Cement, Oil Refining

SÃO PAULO–SANTOS
Food Processing, Textiles,
Chemicals, Iron & Steel,
Machinery, Motor Vehicles,
Oil Refining

LIMA–CALLAO
Textiles, Chemicals,
Leather Goods

CÓRDOBA
Automobiles, Aircraft,
Food Processing,
Chemicals, Cement

SANTIAGO–VALPARAÍSO
Textiles, Chemicals, Food
Processing, Metal Products,
Oil Refining, Leather Goods

CONCEPCIÓN
Iron & Steel, Food
Processing, Textiles,
Oil Refining

BUENOS AIRES–ROSARIO
Food Processing, Textiles,
Machinery, Shipbuilding,
Oil Refining, Chemicals

MAJOR MINERAL OCCURRENCES

Al Bauxite
Ag Silver
Au Gold
Be Beryl
C Coal
Cr Chromium
Cu Copper
D Diamonds
Em Emeralds
Fe Iron Ore
G Natural Gas
Hg Mercury
Id Iodine
Mi Mica
Mn Manganese
Mo Molybdenum
N Nitrates
Na Salt
Ni Nickel
O Petroleum
P Phosphates
Pb Lead
Pt Platinum
Q Quartz Crystal
S Sulfur
Sb Antimony
Sn Tin
U Uranium
V Vanadium
W Tungsten
Zn Zinc

⚡ Water Power
▨ Major Industrial Areas

DOMINANT LAND USE

Wheat, Livestock
Wheat, Corn, Livestock
Cereals, Livestock
Diversified Tropical Crops (chiefly plantation agriculture)
Truck Farming, Horticulture, Special Crops
Upland Cultivated Areas
Intensive Livestock Ranching
Upland Livestock Grazing, Limited Agriculture
Extensive Livestock Ranching
Forests
Nonagricultural Land

15

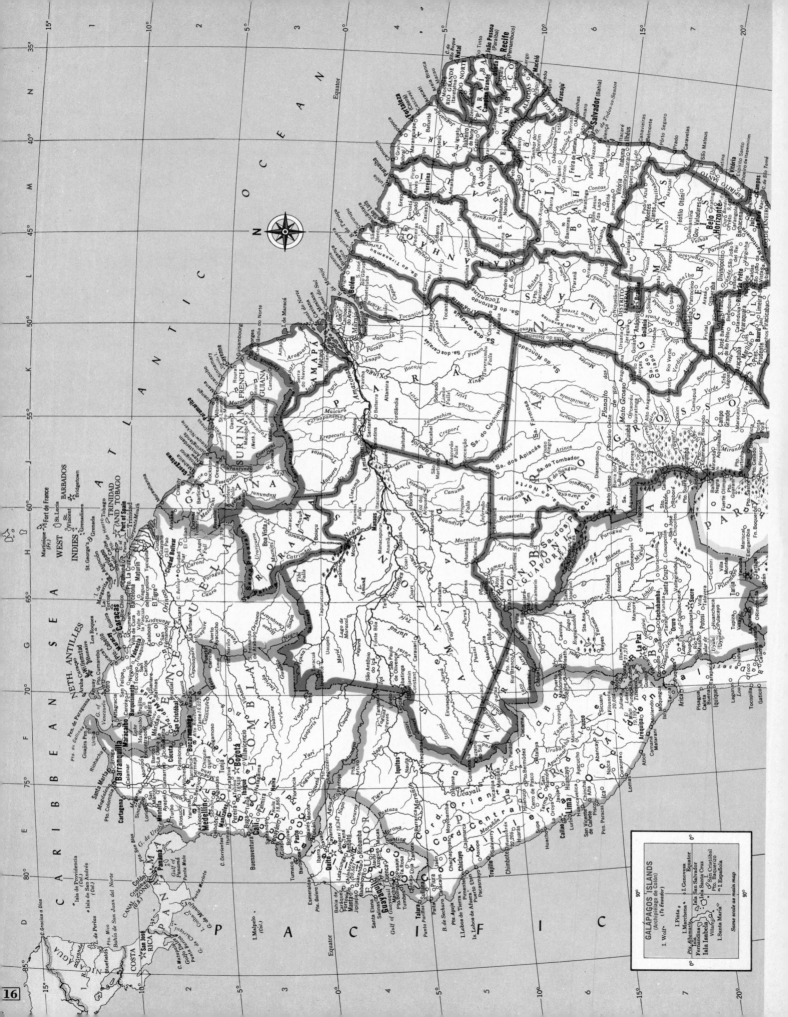

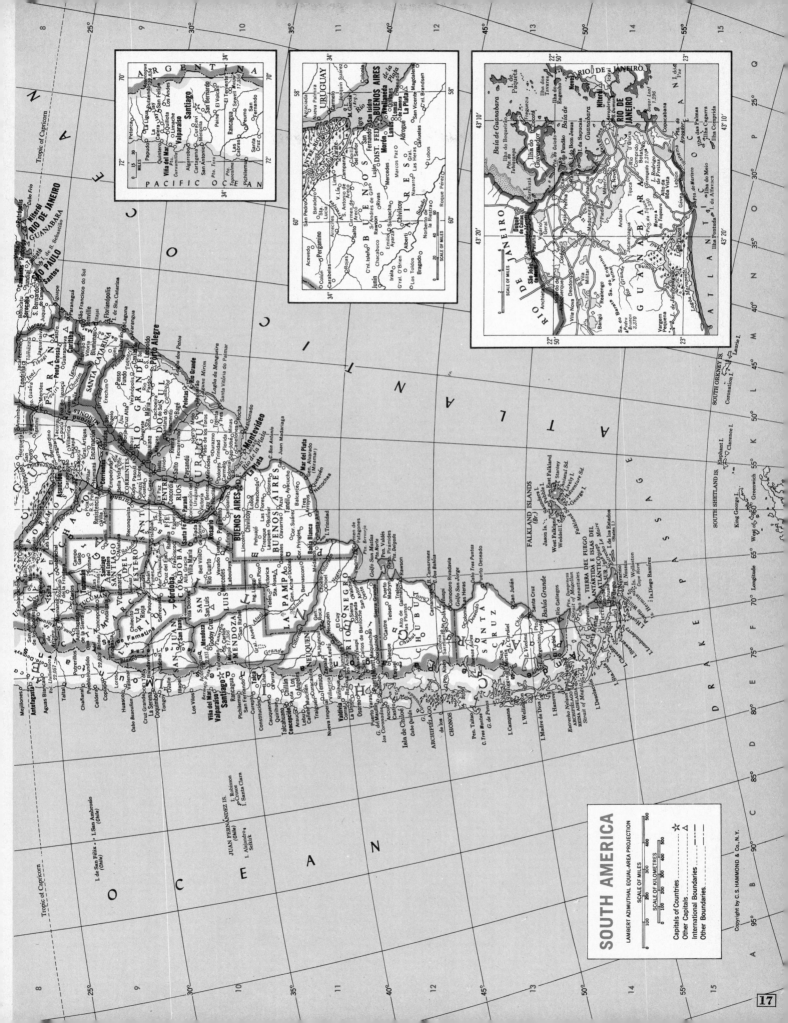

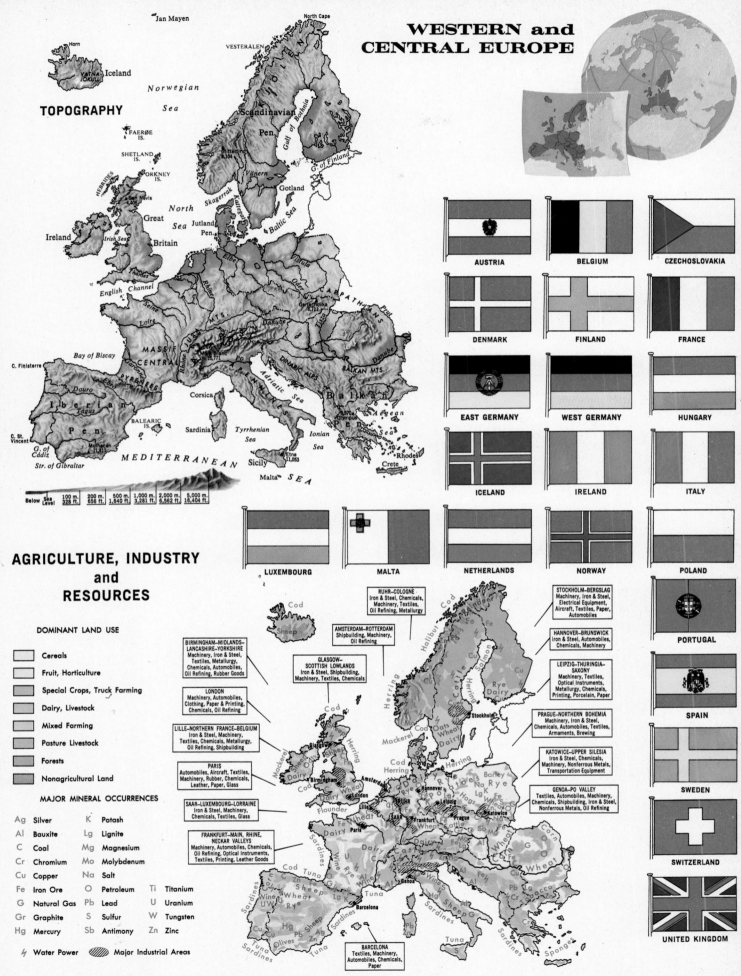

THE BRITISH ISLES

BONNE PROJECTION

SCALE OF MILES

SCALE OF KILOMETRES

Capitals of Countries .... ☆
International Boundaries ....
Other Boundaries ....
Canals ....

SHETLAND ISLANDS

Same scale as main map.

GREATER LONDON

Copyright by C.S. HAMMOND & CO., N.Y.

SVALBARD

NORWEGIAN SEA

SCANDINAVIA
and FINLAND

CONIC PROJECTION

SCALE OF MILES

SCALE OF KILOMETRES

Capitals of Countries .... ☆
Administrative Centers .... △
International Boundaries ....
Internal Boundaries ....
Canals ....

SUBDIVISIONS
Indicated by Numbers
Fylker in NORWAY
1 Akershus G6
2 Vestfold G7
3 Østfold G7
4 Oslo G7
5 Bergen D6
Oslo is the administrative
center for Akershus and
Oslo Fylker; Bergen for
Hordaland and Bergen
Fylker.
Län in SWEDEN
6 Göteborg och G7
  Bohus
7 Västmanland K7
8 Södermanland K7
9 Östergötland J7
10 Malmöhus H9
11 Kristianstad J8

23

ITALY
CONIC PROJECTION

SCALE OF MILES

SCALE OF KILOMETERS

Capitals of Countries	☆
Regional Capitals	⊞
Provincial Capitals	△
International Boundaries	—··—
Regional Boundaries	—·⊕·—

ITALY is divided for administrative purposes into 20 regions, shown on the map in separate colors. The regions of Friuli-Venezia Giulia, Sardinia, Sicily, Trentino-Alto Adige and Valle d'Aosta enjoy special autonomy.

The regions are subdivided into provinces bearing the same names as their respective capitals, except:

PROVINCE	CAPITAL
MASSA-CARRARA	Massa
PESARO-URBINO	Pesaro

VATICAN CITY

ROME and ENVIRONS

Copyright by C.S. HAMMOND & Co., N.Y.

Longitude 12° East of 14° Greenwich

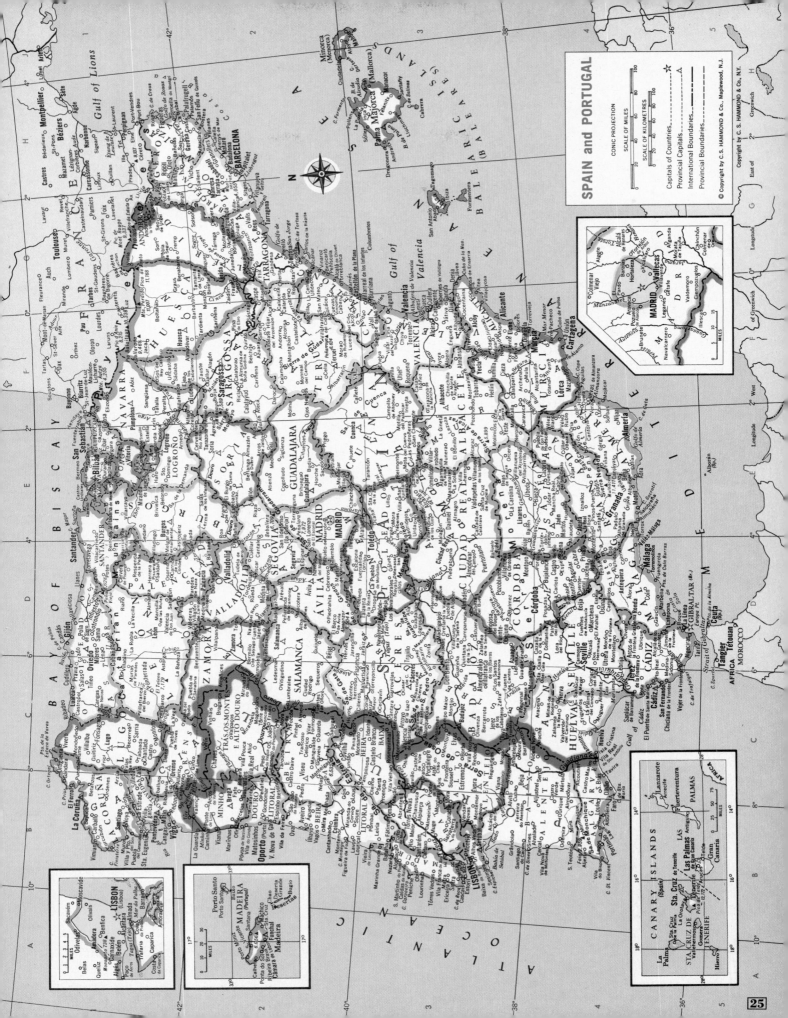

SPAIN and PORTUGAL

CONIC PROJECTION

SCALE OF MILES

SCALE OF KILOMETRES

Capitals of Countries
Provincial Capitals
International Boundaries
Provincial Boundaries

MADRID

CANARY ISLANDS
(Spain)

TENERIFE

PALMAS

LISBON

MADEIRA
(Portugal)

25

THE BALKAN STATES

CONIC PROJECTION

SCALE OF MILES

0   25   50   75   100   125   150   175

SCALE OF KILOMETRES

0   25   50   75   100   125   150   175

Capitals of Countries ⎯⎯⎯⎯⎯ ☆
Administrative Centers ⎯⎯⎯⎯⎯ △
International Boundaries ⎯⎯ ⎯⎯
Major Internal Boundaries ⎯ ⎯ ⎯
Minor Internal Boundaries ⎯ ⋯ ⎯
Canals ⎯⎯⎯⎯⎯⎯⎯⎯⎯⎯

BULGARIA and GREECE are divided into counties and departments, respectively. Because of the scale no attempt has been made to delimit and name these sub-divisions; their administrative centers have, however, been designated.
The larger divisions named in Greece are well-known geographical regions, without administrative function.
RUMANIA consists of thirty-nine counties and three cities of regional status, Bucharest, Constanța and Petroșeni. Scale does not permit delimiting these counties.
ALBANIA is divided into twenty-seven districts. Scale does not permit the delimitation of these divisions.
YUGOSLAVIA is a federation of six republics. The Serbian republic includes an autonomous province (Voyvodina), and an autonomous region (Kosovo-Mitohiyan).

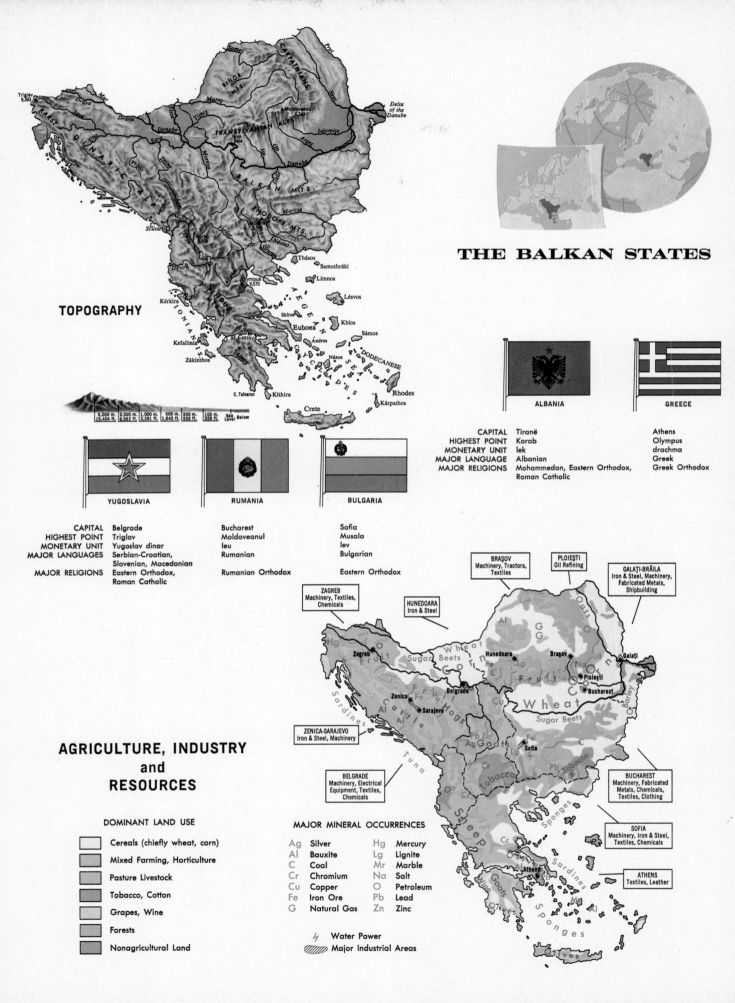

THE BALKAN STATES

TOPOGRAPHY

Triglav 9,393

KARST

DINARIC ALPS

Drava

Sava

Drina

Danube

Mur

Drava

Timis

BIHOR MTS.

CARPATHIANS

Somes

Mures

Prut

Siret

Moldoveanul 9,345

TRANSYLVANIAN ALPS

Iron Gate

Ialomița

Arges

Jiu

Olt

Danube

Delta of the Danube

BALKAN MTS.

RHODOPE MTS.

Maritsa

Morava

Vardar

Struma

Nestos

Scutari

Korab 9,068

Devoll

Mussala 9,596

Olympus 9,570

Thásos

Samothráki

Límnos

Lésvos

Kérkira

IONIAN IS.

Kefallinía

Zákinthos

G. of Corinth

Euboea

Skíros

Ándros

Khíos

Sámos

CYCLADES

AEGEAN SEA

Náxos

DODECANESE

C. Taínaron

Kíthira

Crete

Rhodes

Kárpathos

5,000 m. 16,404 ft. | 2,000 m. 6,562 ft. | 1,000 m. 3,281 ft. | 500 m. 1,640 ft. | 200 m. 656 ft. | 100 m. 328 ft. | Sea Level | Below

ALBANIA

GREECE

CAPITAL	Tiranë	Athens
HIGHEST POINT	Korab	Olympus
MONETARY UNIT	lek	drachma
MAJOR LANGUAGE	Albanian	Greek
MAJOR RELIGIONS	Mohammedan, Eastern Orthodox, Roman Catholic	Greek Orthodox

YUGOSLAVIA

RUMANIA

BULGARIA

CAPITAL	Belgrade	Bucharest	Sofia
HIGHEST POINT	Triglav	Moldoveanul	Musala
MONETARY UNIT	Yugoslav dinar	leu	lev
MAJOR LANGUAGES	Serbian-Croatian, Slovenian, Macedonian	Rumanian	Bulgarian
MAJOR RELIGIONS	Eastern Orthodox, Roman Catholic	Rumanian Orthodox	Eastern Orthodox

AGRICULTURE, INDUSTRY and RESOURCES

ZAGREB
Machinery, Textiles, Chemicals

HUNEDOARA
Iron & Steel

BRAŞOV
Machinery, Tractors, Textiles

PLOIEŞTI
Oil Refining

GALAȚI-BRĂILA
Iron & Steel, Machinery, Fabricated Metals, Shipbuilding

ZENICA-SARAJEVO
Iron & Steel, Machinery

BELGRADE
Machinery, Electrical Equipment, Textiles, Chemicals

BUCHAREST
Machinery, Fabricated Metals, Chemicals, Textiles, Clothing

SOFIA
Machinery, Iron & Steel, Textiles, Chemicals

ATHENS
Textiles, Leather

Zagreb

Zenica

Sarajevo

Belgrade

Hunedoara

Braşov

Ploieşti

Bucharest

Galați

Sofia

Athens

Wheat

Corn

Sugar Beets

Fruit

Cattle

Hogs

Tobacco

Sheep

Goats

Olives

Wine

Sardines

Sponges

Tuna

DOMINANT LAND USE

- Cereals (chiefly wheat, corn)
- Mixed Farming, Horticulture
- Pasture Livestock
- Tobacco, Cotton
- Grapes, Wine
- Forests
- Nonagricultural Land

MAJOR MINERAL OCCURRENCES

Ag	Silver	Hg	Mercury
Al	Bauxite	Lg	Lignite
C	Coal	Mr	Marble
Cr	Chromium	Na	Salt
Cu	Copper	O	Petroleum
Fe	Iron Ore	Pb	Lead
G	Natural Gas	Zn	Zinc

⚡ Water Power

▨ Major Industrial Areas

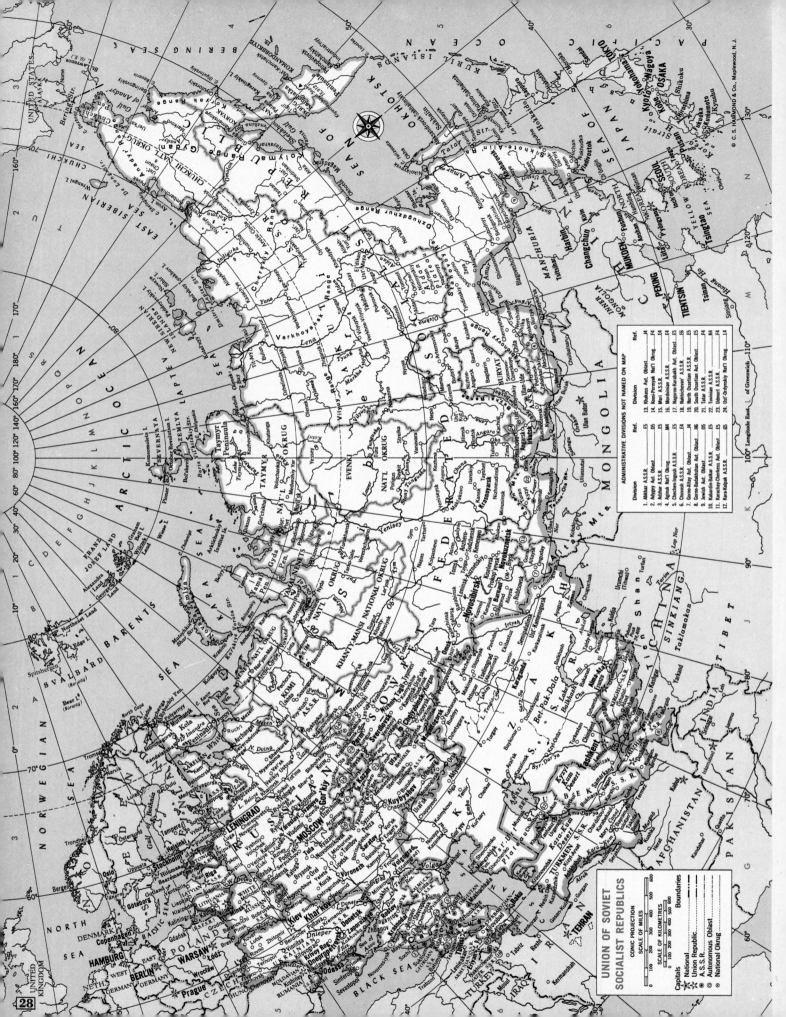

UNION OF SOVIET
SOCIALIST REPUBLICS

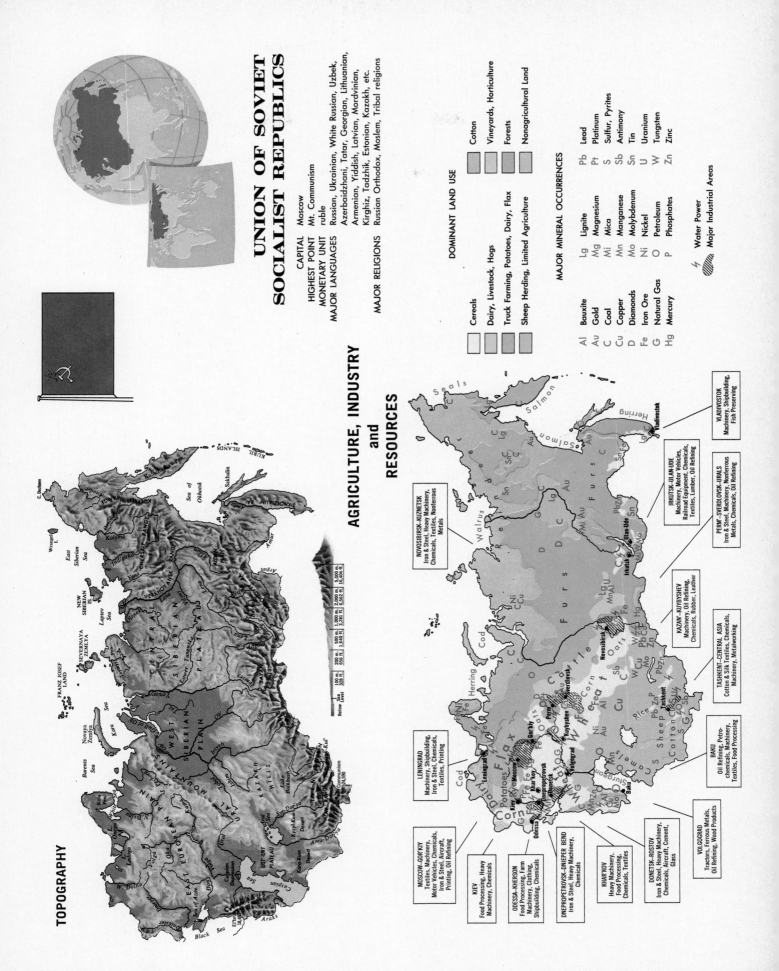

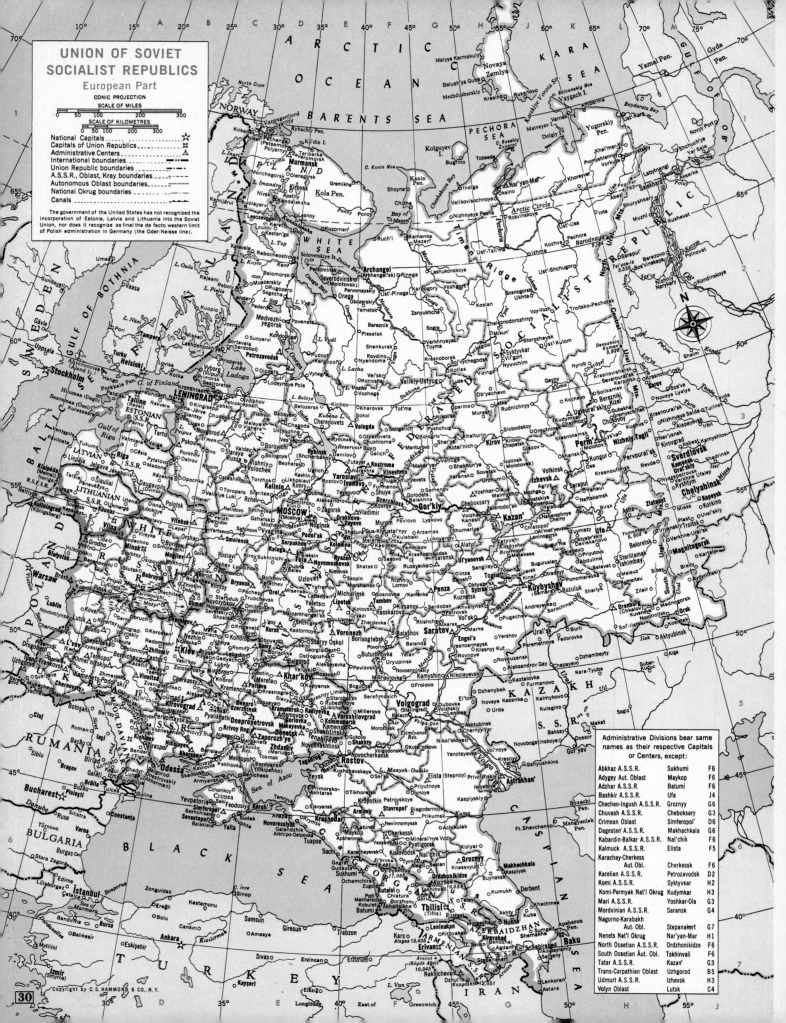

UNION OF SOVIET SOCIALIST REPUBLICS

European Part

CONIC PROJECTION

SCALE OF MILES

SCALE OF KILOMETRES

National Capitals	☆
Capitals of Union Republics	△
Administrative Centers	⊙
International boundaries	
Union Republic boundaries	
A.S.S.R., Oblast, Kray boundaries	
Autonomous Oblast boundaries	
National Okrug boundaries	
Canals	

The government of the United States has not recognized the incorporation of Estonia, Latvia and Lithuania into the Soviet Union, nor does it recognize as final the de facto western limit of Polish administration in Germany (the Oder-Neisse line).

Administrative Divisions bear same names as their respective Capitals or Centers, except:

Abkhaz A.S.S.R.	Sukhumi	F6
Adygey Aut. Oblast	Maykop	F6
Adzhar A.S.S.R.	Batumi	F6
Bashkir A.S.S.R.	Ufa	J4
Chechen-Ingush A.S.S.R.	Groznyy	G6
Chuvash A.S.S.R.	Cheboksary	G3
Crimean Oblast	Simferopol'	D6
Dagestan A.S.S.R.	Makhachkala	G6
Kabardin-Balkar A.S.S.R.	Nal'chik	F6
Kalmuck A.S.S.R.	Elista	F5
Karachay-Cherkess Aut. Obl.	Cherkessk	F6
Karelian A.S.S.R.	Petrozavodsk	D2
Komi A.S.S.R.	Syktyvkar	H2
Komi-Permyak Nat'l Okrug	Kudymkar	H3
Mari A.S.S.R.	Yoshkar-Ola	G3
Mordvinian A.S.S.R.	Saransk	G4
Nagorno-Karabakh Aut. Obl.	Stepanakert	G7
Nenets Nat'l Okrug	Nar'yan-Mar	H1
North Ossetian A.S.S.R.	Ordzhonikidze	F6
South Ossetian Aut. Obl.	Tskhinvali	F6
Tatar A.S.S.R.	Kazan'	G3
Trans-Carpathian Oblast	Uzhgorod	B5
Udmurt A.S.S.R.	Izhevsk	H3
Volyn Oblast	Lutsk	C4

Copyright by C.S. HAMMOND & CO., N.Y.

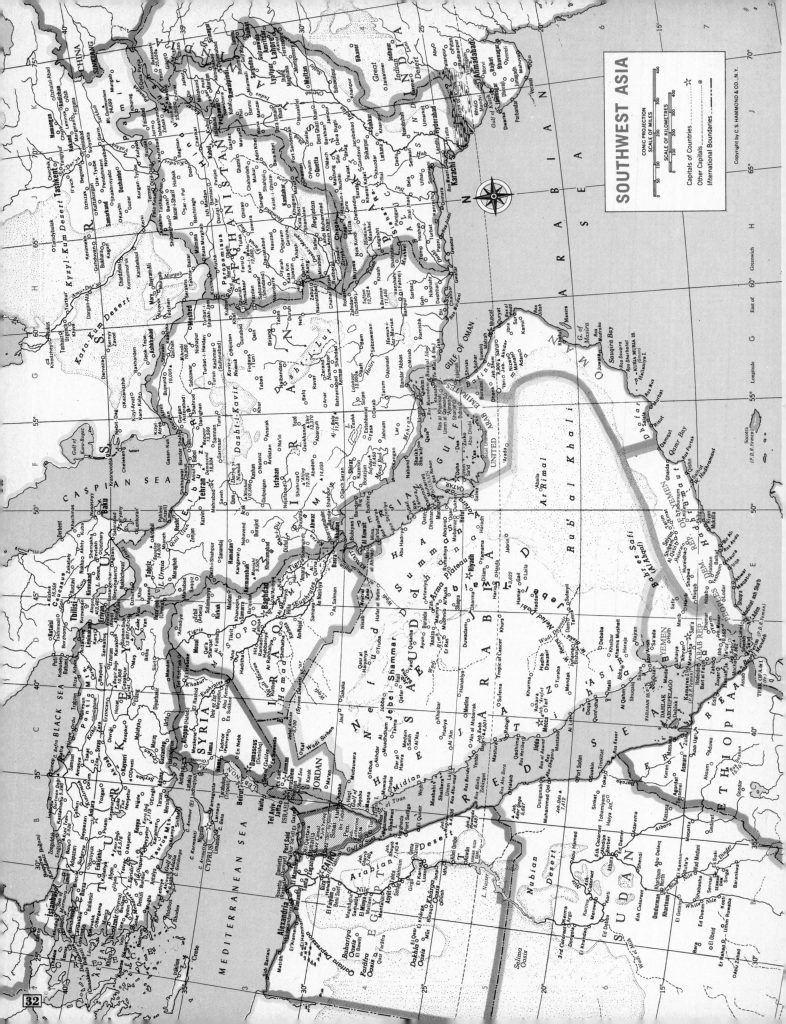

SOUTHWEST ASIA

CONIC PROJECTION
SCALE OF MILES
SCALE OF KILOMETRES

Capitals of Countries .... ★
Other Capitals .... ◉
International Boundaries .... ----

Copyright by C.S. HAMMOND & CO., N.Y.

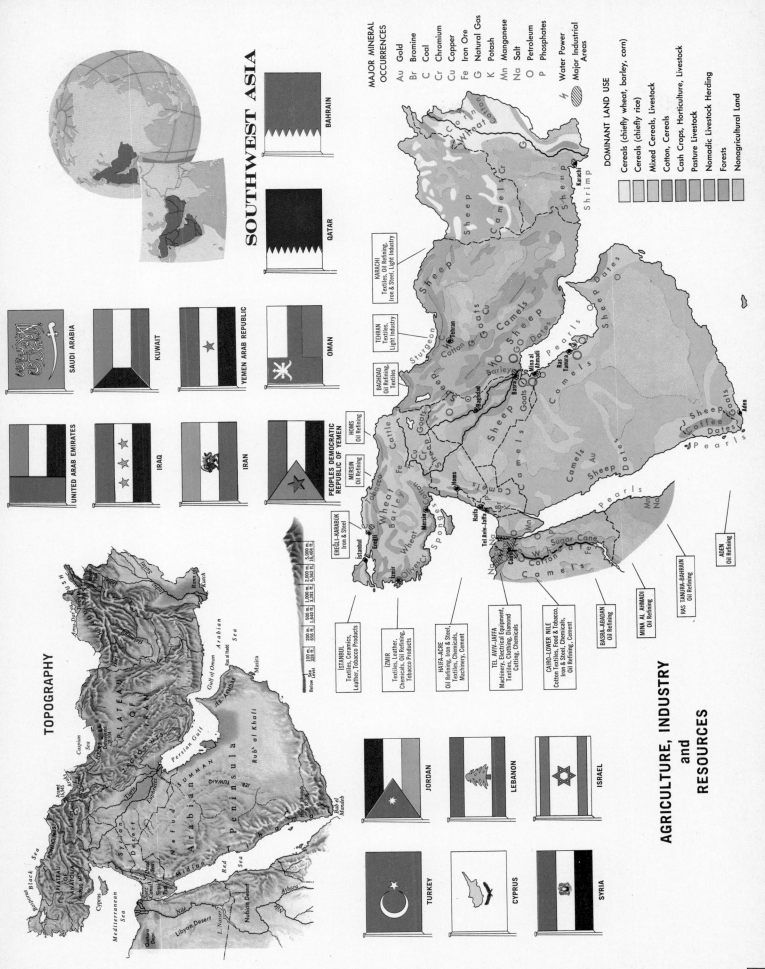

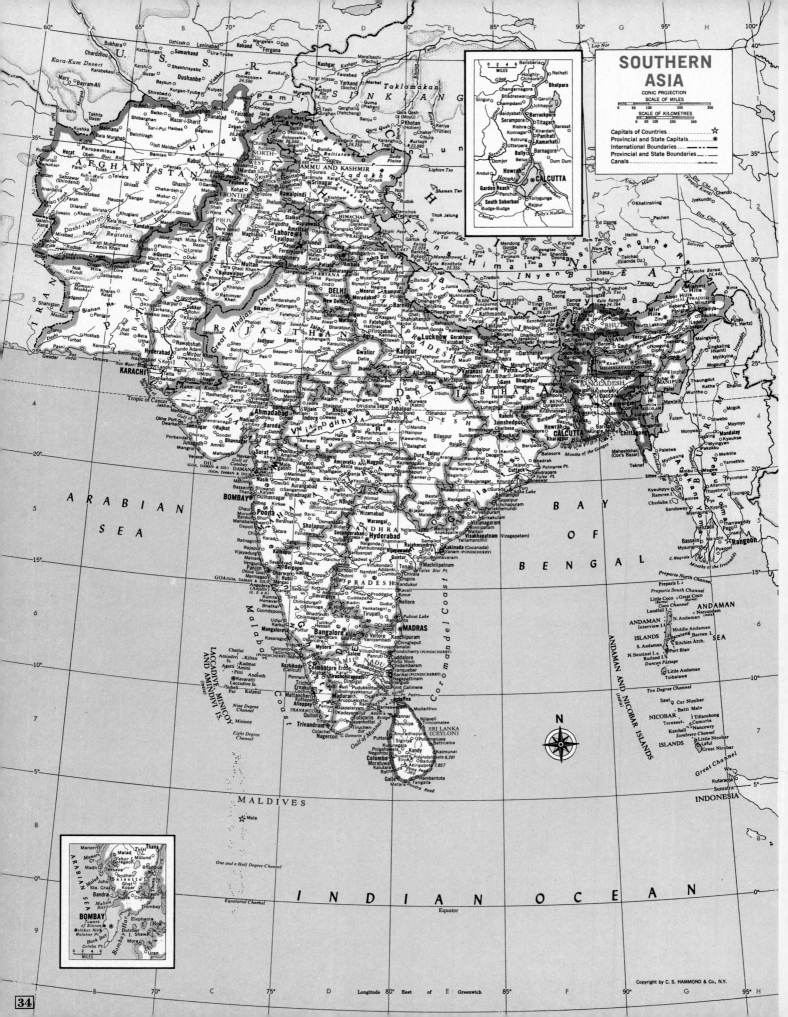

SOUTHERN ASIA

CONIC PROJECTION

SCALE OF MILES

SCALE OF KILOMETRES

Capitals of Countries .... ☆
Provincial and State Capitals .... ◉
International Boundaries ....
Provincial and State Boundaries ....
Canals ....

Copyright by C. S. HAMMOND & Co., N.Y.

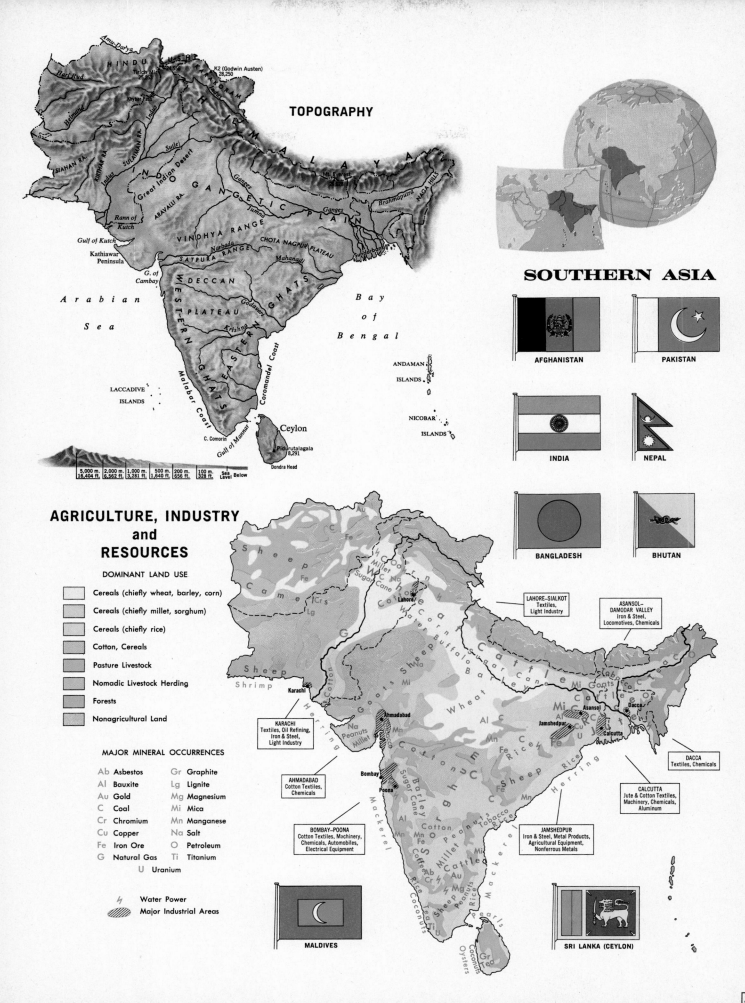

EAST ASIA

CONIC PROJECTION

SCALE OF MILES

SCALE OF KILOMETRES

Capitals of Countries..........
Provincial Capitals...........
Canals.....................

International Boundaries......
Provincial Boundaries.......
Walls....................

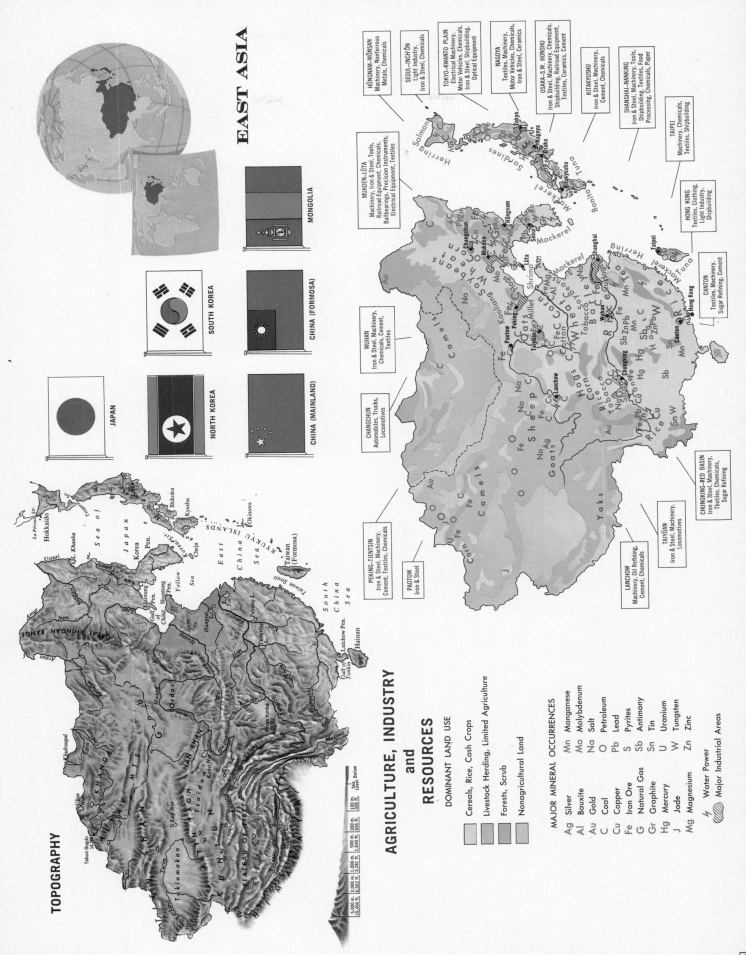

EAST ASIA

TOPOGRAPHY

AGRICULTURE, INDUSTRY and RESOURCES

37

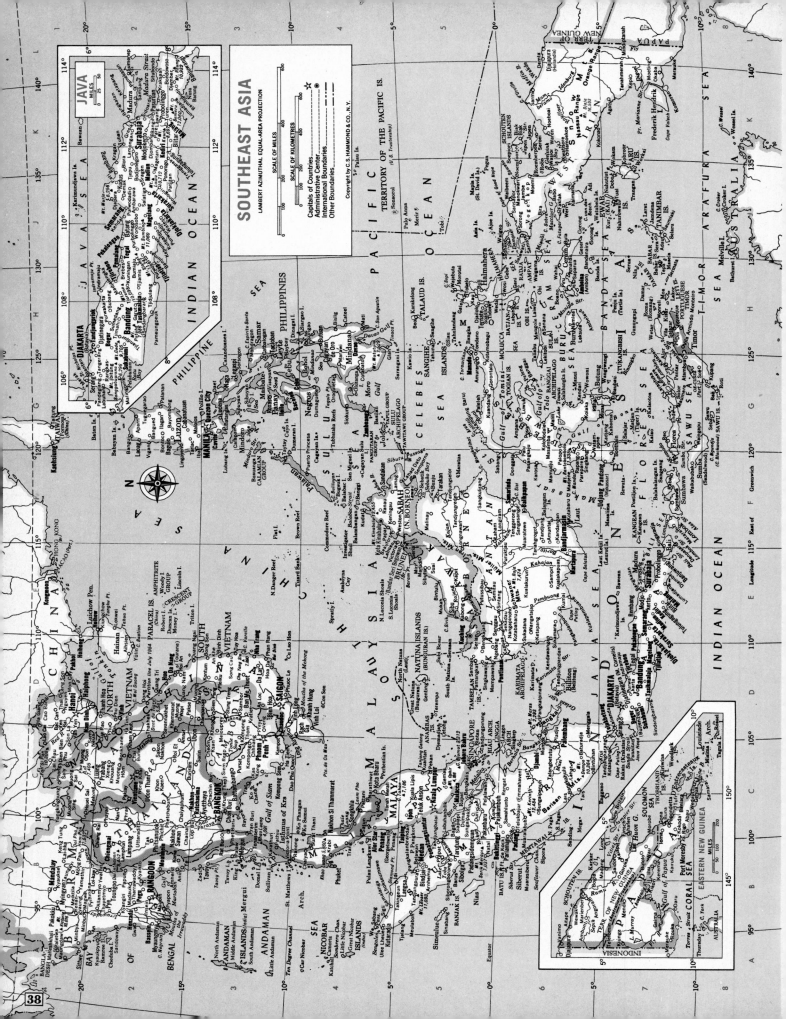

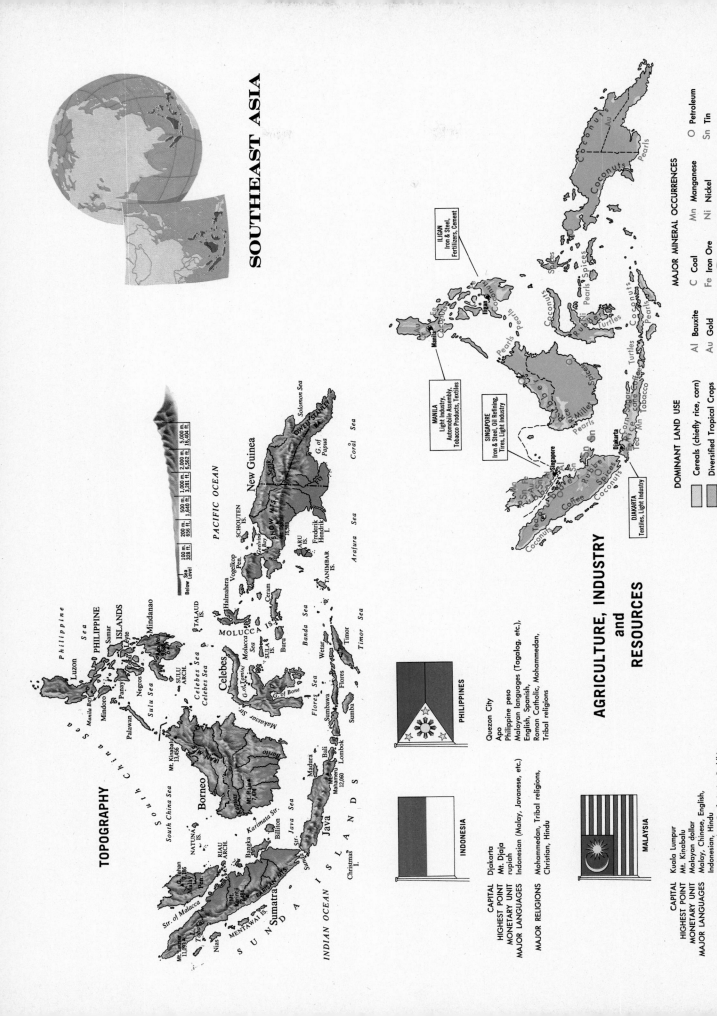

SOUTHEAST ASIA

TOPOGRAPHY

Below Sea Level | Sea Level | 100 m. 328 ft. | 200 m. 656 ft. | 500 m. 1,640 ft. | 1,000 m. 3,281 ft. | 2,000 m. 6,562 ft. | 5,000 m. 16,404 ft.

INDONESIA

CAPITAL   Djakarta
HIGHEST POINT   Mt. Djaja
MONETARY UNIT   rupiah
MAJOR LANGUAGES   Indonesian (Malay, Javanese, etc.)
MAJOR RELIGIONS   Mohammedan, Tribal religions, Christian, Hindu

PHILIPPINES

   Quezon City
   Apo
   Philippine peso
   Malayan languages (Tagalog, etc.), English, Spanish, Roman Catholic, Mohammedan, Tribal religions

MALAYSIA

CAPITAL   Kuala Lumpur
HIGHEST POINT   Mt. Kinabalu
MONETARY UNIT   Malayan dollar
MAJOR LANGUAGES   Malay, Chinese, English, Indonesian, Hindu
MAJOR RELIGIONS   Mohammedan, Confucianist, Buddhist, Tribal religions, Hindu, Taoist

AGRICULTURE, INDUSTRY and RESOURCES

ILIGAN
Iron & Steel, Fertilizers, Cement

MANILA
Light Industry, Automobile Assembly, Tobacco Products, Textiles

SINGAPORE
Iron & Steel, Oil Refining, Tires, Light Industry

DJAKARTA
Textiles, Light Industry

DOMINANT LAND USE

Cereals (chiefly rice, corn)
Diversified Tropical Crops
Forests

MAJOR MINERAL OCCURRENCES

C Coal   Mn Manganese   O Petroleum
Fe Iron Ore   Ni Nickel   Sn Tin
Al Bauxite   Au Gold   ▨ Major Industrial Areas

INDOCHINESE
and
MALAY PENINSULAS

CONIC PROJECTION

SCALE OF MILES

SCALE OF KILOMETRES

International Boundaries
Division and State Boundaries
Capitals of Countries
Division and State Capitals

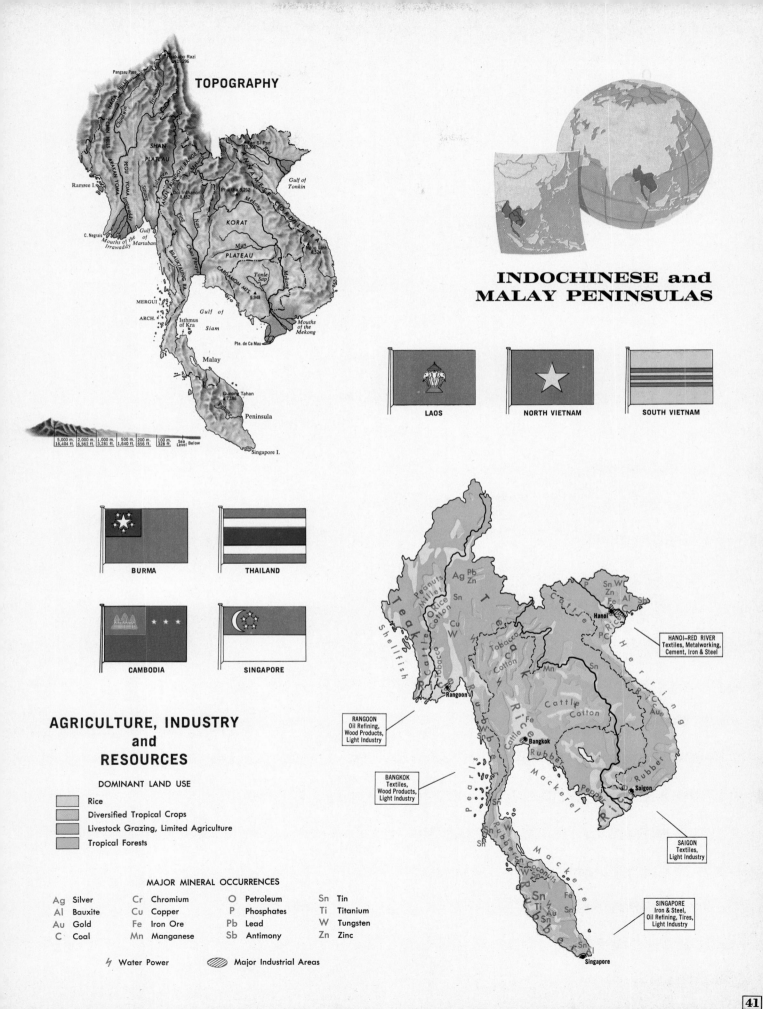

TOPOGRAPHY

Hkakabo Razi
19,296
Pangsau Pass
NAGA HILLS
CHIN HILLS
ARAKAN YOMA
PEGU YOMA
Chindwin
Irrawaddy
SHAN PLATEAU
Salween
Sittang
TANEN-TAUNGGYI RANGE
Maung
Ping
Nan
Doi Inthanon 8,452
Fan Si Pan 10,308
Red
Gulf of Tonkin
Chao Praya
Mekong
Phu Bia 9,252
KORAT PLATEAU
Mun
ANNAMESE CORDILLERA
Ngoc Linh 8,524
Ramree I.
C. Negrais
Mouths of the Irrawaddy
Gulf of Martaban
BILAUKTAUNG RA.
CARDAMOM MTS.
Tonle Sap
5,948
Mouths of the Mekong
MERGUI ARCH.
Isthmus of Kra
Gulf of Siam
Pte. de Ca Mau
Malay
Gunong Tahan 7,186
Pahang
Peninsula
Singapore I.

| 5,000 m. 16,404 ft. | 2,000 m. 6,562 ft. | 1,000 m. 3,281 ft. | 500 m. 1,640 ft. | 200 m. 656 ft. | 100 m. 328 ft. | Sea Level | Below |

INDOCHINESE and MALAY PENINSULAS

LAOS

NORTH VIETNAM

SOUTH VIETNAM

BURMA

THAILAND

CAMBODIA

SINGAPORE

AGRICULTURE, INDUSTRY and RESOURCES

DOMINANT LAND USE

- Rice
- Diversified Tropical Crops
- Livestock Grazing, Limited Agriculture
- Tropical Forests

MAJOR MINERAL OCCURRENCES

Ag	Silver	Cr	Chromium	O	Petroleum	Sn	Tin
Al	Bauxite	Cu	Copper	P	Phosphates	Ti	Titanium
Au	Gold	Fe	Iron Ore	Pb	Lead	W	Tungsten
C	Coal	Mn	Manganese	Sb	Antimony	Zn	Zinc

⚡ Water Power   ▨ Major Industrial Areas

HANOI—RED RIVER
Textiles, Metalworking, Cement, Iron & Steel

RANGOON
Oil Refining, Wood Products, Light Industry

BANGKOK
Textiles, Wood Products, Light Industry

SAIGON
Textiles, Light Industry

SINGAPORE
Iron & Steel, Oil Refining, Tires, Light Industry

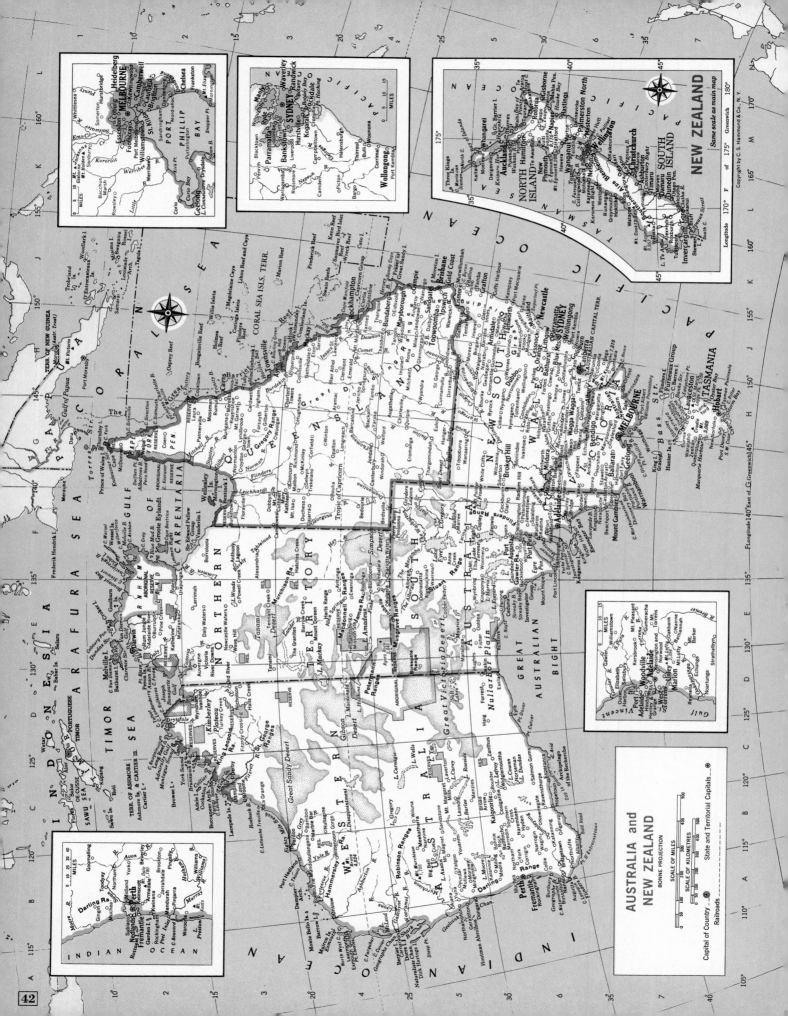

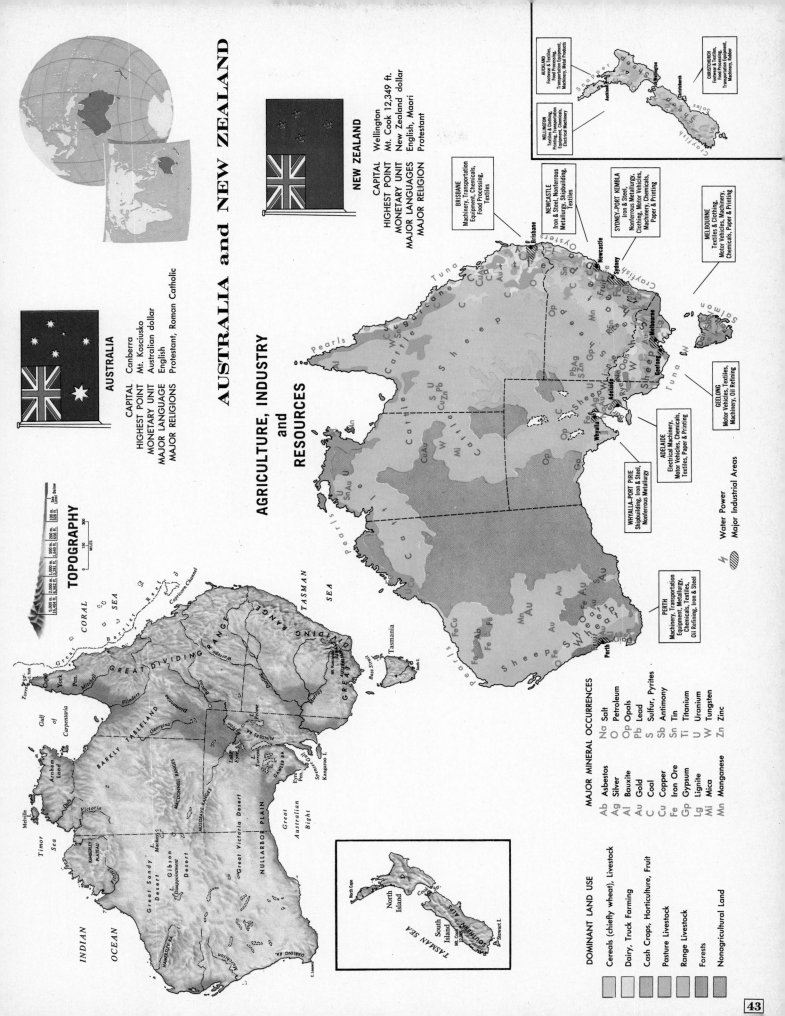

AUSTRALIA and NEW ZEALAND

AGRICULTURE, INDUSTRY and RESOURCES

TOPOGRAPHY

AUSTRALIA

CAPITAL	Canberra
HIGHEST POINT	Mt. Kosciusko
MONETARY UNIT	Australian dollar
MAJOR LANGUAGE	English
MAJOR RELIGIONS	Protestant, Roman Catholic

NEW ZEALAND

CAPITAL	Wellington
HIGHEST POINT	Mt. Cook 12,349 ft.
MONETARY UNIT	New Zealand dollar
MAJOR LANGUAGES	English, Maori
MAJOR RELIGION	Protestant

DOMINANT LAND USE

- Cereals (chiefly wheat), Livestock
- Dairy, Truck Farming
- Cash Crops, Horticulture, Fruit
- Pasture Livestock
- Range Livestock
- Forests
- Nonagricultural Land

MAJOR MINERAL OCCURRENCES

Ab	Asbestos	Na	Salt
Ag	Silver	O	Petroleum
Al	Bauxite	Op	Opals
Au	Gold	Pb	Lead
C	Coal	S	Sulfur, Pyrites
Cu	Copper	Sb	Antimony
Fe	Iron Ore	Sn	Tin
Gp	Gypsum	Ti	Titanium
Lg	Lignite	U	Uranium
Mi	Mica	W	Tungsten
Mn	Manganese	Zn	Zinc

Water Power

Major Industrial Areas

BRISBANE Machinery, Transportation Equipment, Food Processing, Textiles

NEWCASTLE Iron & Steel, Nonferrous Metallurgy, Shipbuilding, Textiles

SYDNEY–PORT KEMBLA Iron & Steel, Nonferrous Metallurgy, Clothing, Motor Vehicles, Machinery, Chemicals, Paper & Printing

MELBOURNE Textiles & Clothing, Motor Vehicles, Machinery, Chemicals, Paper & Printing

GEELONG Motor Vehicles, Textiles, Machinery, Oil Refining

ADELAIDE Electrical Machinery, Motor Vehicles, Chemicals, Textiles, Paper & Printing

WHYALLA–PORT PIRIE Shipbuilding, Iron & Steel, Nonferrous Metallurgy

PERTH Machinery, Transportation Equipment, Metallurgy, Chemicals, Textiles, Oil Refining, Iron & Steel

AUCKLAND Footwear & Textiles, Food Processing, Transportation Equipment, Machinery, Metal Products

WELLINGTON Textiles & Clothing, Printing, Transportation Equipment, Chemicals, Electrical Machinery

CHRISTCHURCH Footwear & Textiles, Food Processing, Transportation Equipment, Machinery, Rubber

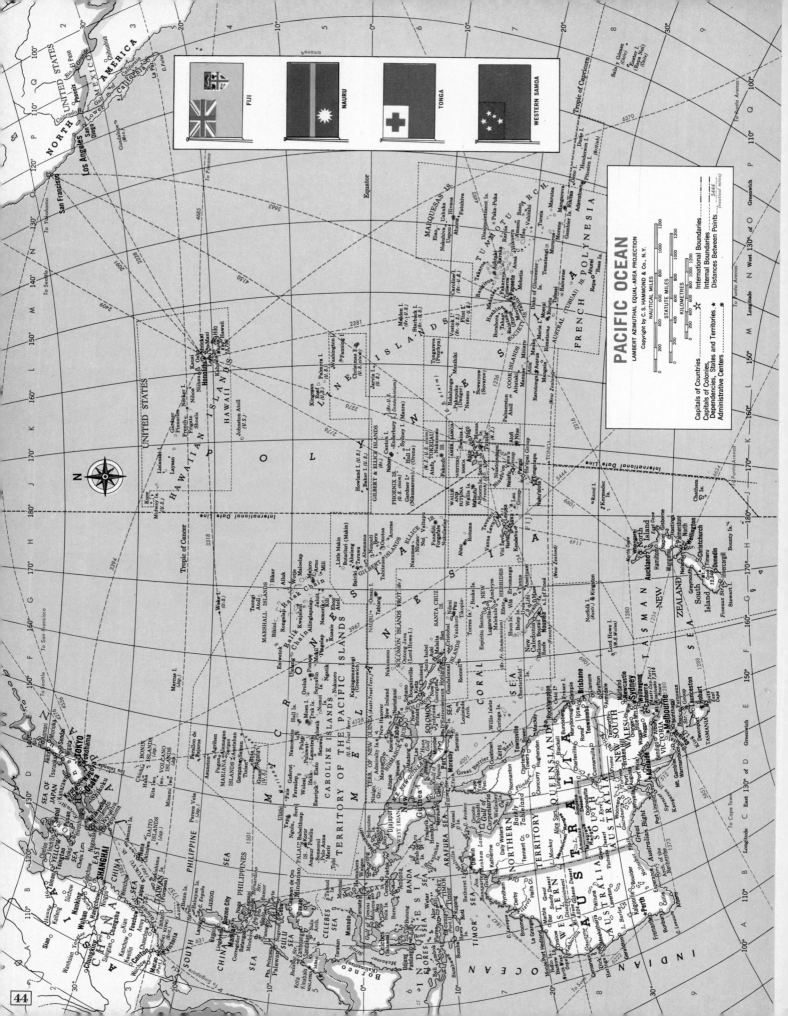

AFRICA

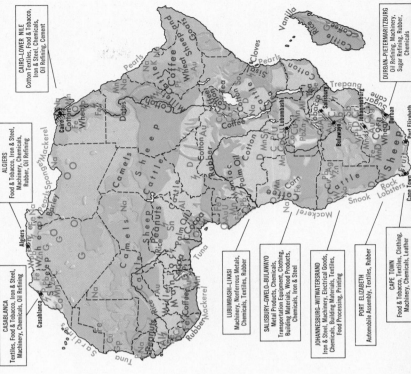

CAIRO–LOWER NILE
Cotton Textiles, Food & Tobacco, Iron & Steel, Chemicals, Oil Refining, Cement

ALGIERS
Food & Tobacco, Iron & Steel, Machinery, Chemicals, Rubber, Oil Refining

CASABLANCA
Textiles, Food & Tobacco, Iron & Steel, Machinery, Chemicals, Oil Refining

LUBUMBASHI–LIKASI
Machinery, Nonferrous Metals, Chemicals, Textiles, Rubber

SALISBURY–GWELO–BULAWAYO
Metal Products, Chemicals, Transportation Equipment, Clothing, Building Materials, Wood Products, Chemicals, Iron & Steel

JOHANNESBURG–WITWATERSRAND
Iron & Steel, Machinery, Electrical Goods, Chemicals, Building Materials, Textiles, Food Processing, Printing

PORT ELIZABETH
Automobile Assembly, Textiles, Rubber

CAPE TOWN
Food & Tobacco, Textiles, Clothing, Machinery, Chemicals, Leather

DURBAN–PIETERMARITZBURG
Oil Refining, Machinery, Sugar Refining, Rubber, Chemicals

TOPOGRAPHY

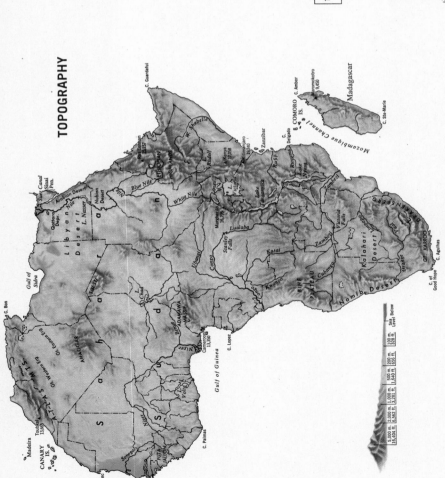

AGRICULTURE, INDUSTRY and RESOURCES

DOMINANT LAND USE

- Cereals, Horticulture, Livestock
- Cash Crops, Mixed Cereals
- Cotton, Cereals
- Diversified Tropical Crops
- Plantation Agriculture
- Oases
- Pasture Livestock
- Nomadic Livestock Herding
- Forests
- Nonagricultural Land

MAJOR MINERAL OCCURRENCES

Ab	Asbestos	Sb	Antimony
Ag	Silver	Sn	Tin
Al	Bauxite	So	Soda Ash
Au	Gold	Ti	Titanium
Be	Beryl	U	Uranium
C	Coal	V	Vanadium
Co	Cobalt	W	Tungsten
Cr	Chromium	Zn	Zinc
Cu	Copper		
D	Diamonds		
Fe	Iron Ore		
G	Natural Gas		
Gp	Gypsum		
Gr	Graphite		
K	Potash		
Mi	Mica		
Mn	Manganese		
Na	Salt		
O	Petroleum		
P	Phosphates		
Pb	Lead		
Pt	Platinum		

Water Power

Major Industrial Areas

45

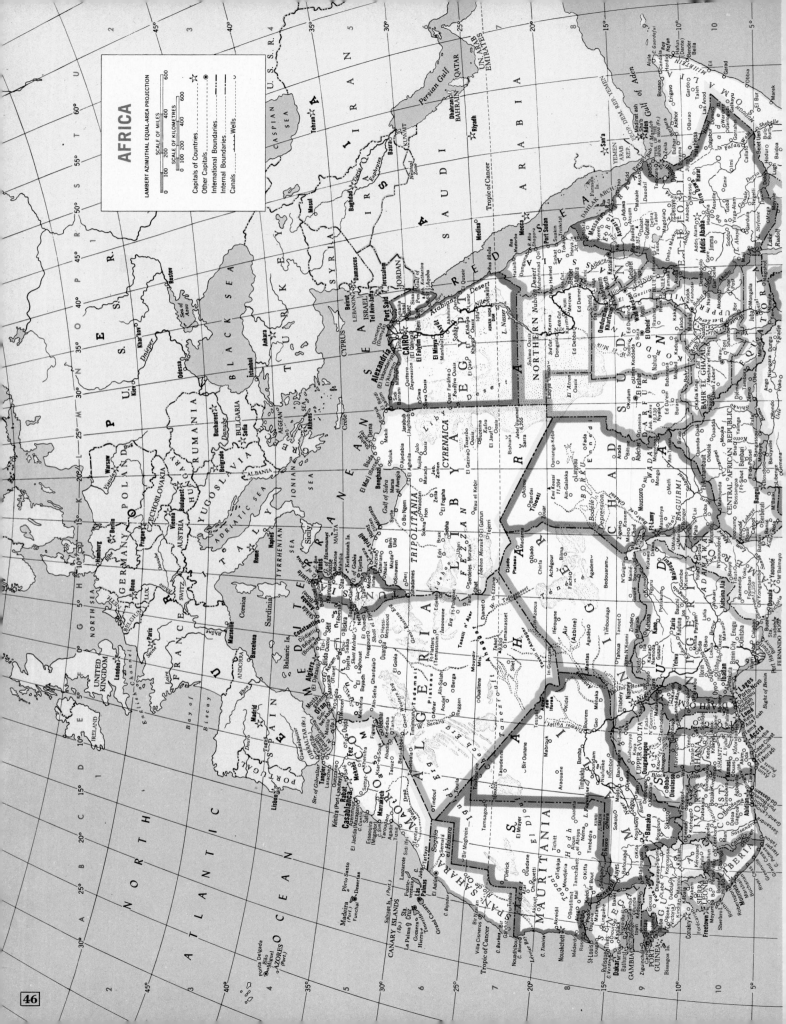

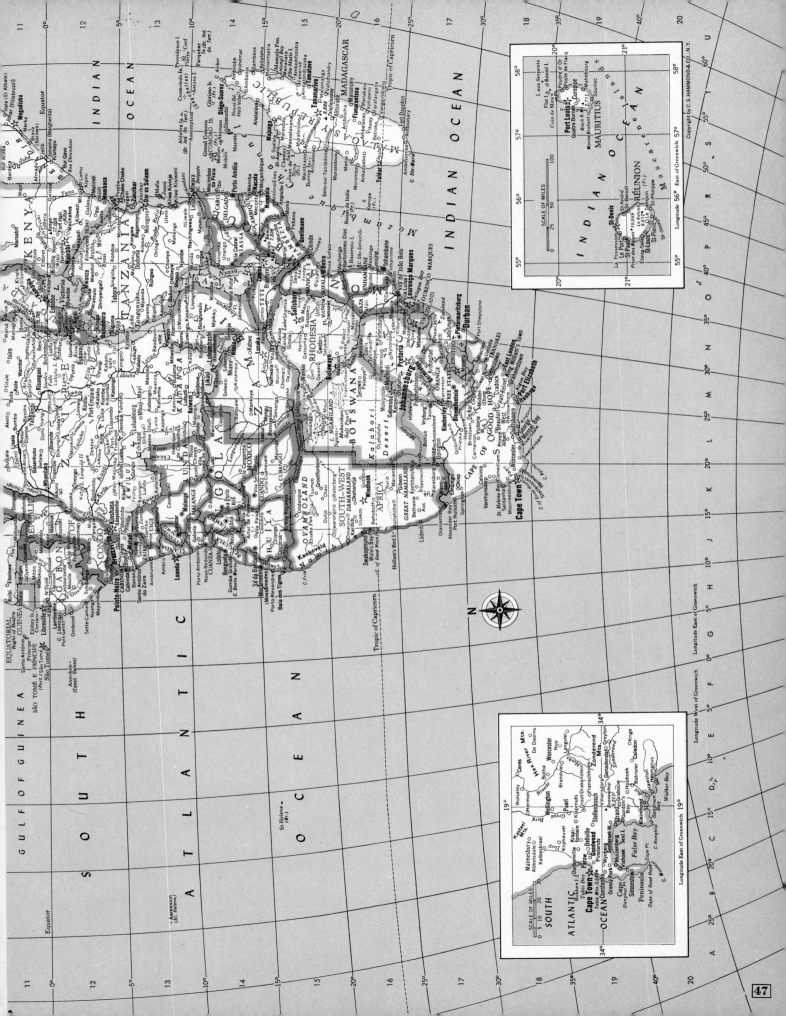

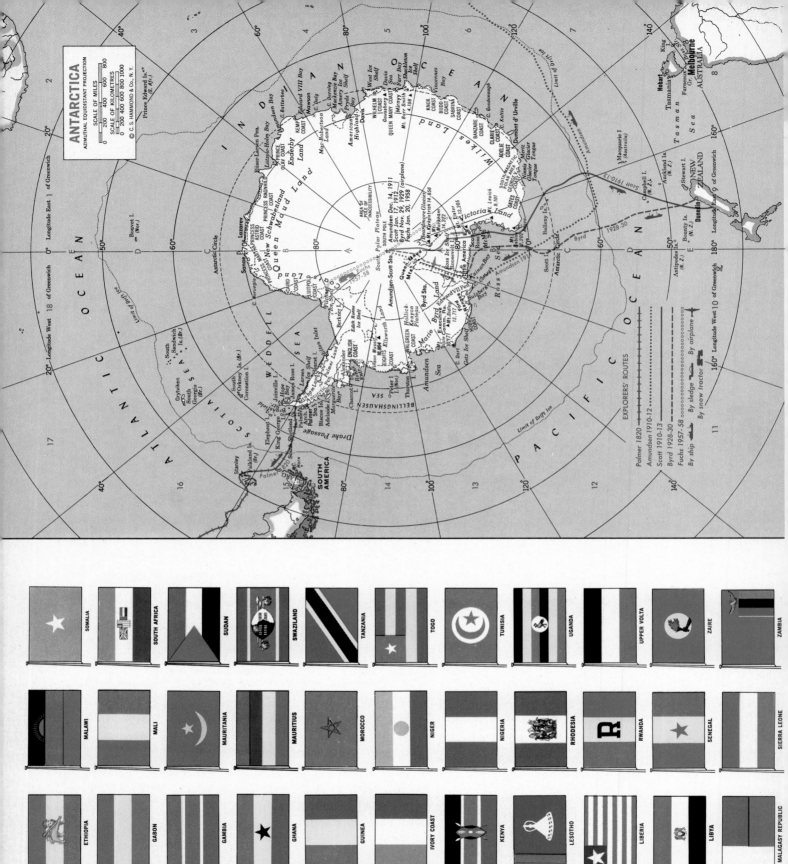

FLAGS OF AFRICA

GAZETTEER-INDEX OF THE WORLD

Country	Area (Sq. Miles)	Population	Capital or Chief Town	Index Ref.	Plate No.
Macao	6.2	292,000	Macao	H 7	36
Maine, U.S.A.	33,215	993,663	Augusta	R 3	7
* Malagasy Republic	226,657	7,011,563	Tananarive	R15	47
* Malawi	45,483	4,530,000	Zomba	N14	47
Malaya, Malaysia	50,670	9,000,000	Kuala Lumpur	D 6	40
* Malaysia	128,308	10,583,000	Kuala Lumpur	B-F4	38
* Maldives	115	110,770	Male	C 8	34
* Mali	463,948	4,929,000	Bamako	E 9	46
* Malta	122	321,000	Valletta	E 7	24
Manitoba, Canada	251,000	979,000	Winnipeg	G 6	3
Martinique	425	332,000	Fort-de-France	G 4	12
Maryland, U.S.A.	10,577	3,922,399	Annapolis	O 5	7
Massachusetts, U.S.A.	8,257	5,689,170	Boston	R 4	7
* Mauritania	397,954	1,140,000	Nouakchott	C 8	46
* Mauritius	787	823,000	Port Louis	S19	47
* Mexico	761,601	48,313,438	Mexico City	......	8
Michigan, U.S.A.	58,216	8,875,083	Lansing	M 4	7
Midway Islands	2	2,356		J 3	44
Minnesota, U.S.A.	84,068	3,805,069	St. Paul	K 3	7
Mississippi, U.S.A.	47,716	2,216,912	Jackson	L 6	7
Missouri, U.S.A.	69,686	4,677,399	Jefferson City	K 5	7
Monaco	368 acres	23,035	Monaco	G 6	21
* Mongolia	604,090	1,300,000	Ulan Bator	F 2	36
Montana, U.S.A.	147,138	694,409	Helena	E 3	6
Montserrat	38	12,300	Plymouth	F 3	12
* Morocco	172,413	15,577,000	Rabat	E 5	46
Mozambique	302,328	7,376,000	Lourenço Marques	N17	47
Nauru	8.2	7,000	Uaboe dist.	G 6	44
Nebraska, U.S.A.	77,227	1,483,791	Lincoln	J 4	6
* Nepal	54,362	10,845,000	Kathmandu (Katmandu)	E 3	34
* Netherlands	13,958	13,077,000	Amsterdam, The Hague	H 5	18
Netherlands Antilles	390	220,000	Willemstad	E 4	12
Nevada, U.S.A.	110,540	488,738	Carson City	D 5	6
New Brunswick, Canada	28,354	624,000	Fredericton	K 6	3
New Caledonia & Dependencies	8,548	100,579	Nouméa	G 8	44
Newfoundland, Canada	156,185	520,000	St. John's	K 4	3
New Guinea, Terr. of (Aust. Trust.)	92,160	1,722,572	Port Moresby	E 6	44
New Hampshire, U.S.A.	9,304	737,681	Concord	P 4	7
New Hebrides	5,700	80,000	Vila	G 7	44
New Jersey, U.S.A.	7,836	7,168,164	Trenton	P 4	7
New Mexico, U.S.A.	121,666	1,016,000	Santa Fe	G 6	6
New York, U.S.A.	49,576	18,190,740	Albany	P 4	7
* New Zealand	103,736	2,815,000	Wellington	M 7	42
* Nicaragua	45,698	1,984,000	Managua	D 4	10
* Niger	489,189	4,016,000	Niamey	G 9	46
* Nigeria	356,669	66,174,000	Lagos	G10	46
Niue	100	5,323	Alofi	K 7	44
North America	9,363,000	314,000,000		......	3
North Carolina, U.S.A.	52,586	5,082,059	Raleigh	O 6	7
North Dakota, U.S.A.	70,665	617,761	Bismarck	H 3	6
Northern Ireland, U.K.	5,459	1,512,500	Belfast	C 3	20
Northwest Territories, Canada	1,304,903	34,000	Yellowknife	E 3	3
* Norway	125,181	3,893,000	Oslo	......	23
Nova Scotia, Canada	21,425	767,000	Halifax	K 6	3
Ohio, U.S.A.	41,222	10,652,017	Columbus	N 5	7
Oklahoma, U.S.A.	69,919	2,559,253	Oklahoma City	J 6	6
* Oman	82,000	565,000	Muscat	G 5	32
Ontario, Canada	412,582	7,707,000	Toronto	H 6	3
Oregon, U.S.A.	96,981	2,091,385	Salem	C 3	6
Pacific Islands, U.S. Trust Terr. of the	687	98,009	Garapan	E 4	44
* Pakistan	310,403	60,000,000	Islamabad	......	34
* Panama	29,209	1,425,343	Panamá	H 6	10
Papua, Australia	86,100	648,000	Port Moresby	B 7	38
* Paraguay	157,047	2,314,000	Asunción	J 9	17
Pennsylvania, U.S.A.	45,333	11,793,909	Harrisburg	O 4	7
* Persia (Iran)	636,293	28,448,000	Tehran	F 2	32
* Peru	496,222	13,586,300	Lima	E 6	16
* Philippines	115,707	39,079,000	Quezon City	G 3	38
Pitcairn Islands	18	80	Adamstown	N 8	44
* Poland	120,702	32,889,000	Warsaw	L 5	18
* Portugal	35,510	9,560,000	Lisbon	......	25
Portuguese Guinea	13,948	530,000	Bissau	C 9	46
Portuguese Timor	5,762	590,000	Dili	H 7	38
Prince Edward Island, Canada	2,184	110,000	Charlottetown	K 6	3
Puerto Rico	3,435	2,689,932	San Juan	G 1	12
* Qatar	8,500	100,000	Doha	F 4	32
Québec, Canada	594,860	6,023,000	Québec	J 6	3
Réunion	970	436,000	St-Denis	P20	47
Rhode Island, U.S.A.	1,214	949,723	Providence	R 4	7
Rhodesia	150,332	5,310,000	Salisbury	N15	47
* Rumania	91,699	20,394,000	Bucharest	G 3	26
* Rwanda	10,169	3,500,000	Kigali	N12	47
Sabah, Malaysia	29,388	633,000	Kota Kinabalu	F 4	38
St. Christopher-Nevis-Anguilla	138	56,000	Basseterre	F 3	12
St. Helena	47	6,462	Jamestown	E15	47
St. Lucia	238	110,000	Castries	G 4	12
St-Pierre and Miquelon	93.5	5,235	St-Pierre	L 6	3
St. Vincent	150	95,000	Kingstown	G 4	12
San Marino	23.4	19,000	San Marino	D 3	24
São Tomé e Príncipe	372	66,000	São Tomé	H11	47
Sarawak, Malaysia	48,250	950,000	Kuching	E 5	38
Saskatchewan, Canada	251,700	933,000	Regina	F 5	3
* Saudi Arabia	920,000	7,200,000	Riyadh, Mecca	D 4	32
Scotland, U.K.	30,411	5,194,000	Edinburgh	D 2	20
* Senegal	75,750	3,780,000	Dakar	C 9	46
Seychelles	91	51,396	Victoria	J10	31
* Siam (Thailand)	198,456	35,448,000	Bangkok	D 4	40
* Sierra Leone	27,925	2,512,000	Freetown	D10	46
* Singapore	225	2,034,000	Singapore	F 6	40
Solomon Islands Prot.	11,500	161,525	Honiara	G 6	44
* Somalia	246,200	2,730,000	Mogadishu	R11	47
* South Africa	471,663	21,282,000	Cape Town, Pretoria	L18	47
South America	6,875,000	186,000,000		......	16,17
South Carolina, U.S.A.	31,055	2,590,516	Columbia	N 6	7
South Dakota, U.S.A.	77,047	666,257	Pierre	H 3	6
South-West Africa	317,838	615,000	Windhoek	K16	47
* Spain	194,896	33,290,000	Madrid	......	25
Spanish Sahara, Spain	102,702	63,000	El Aaiún	D 6	46
* Sri Lanka (Ceylon)	25,332	12,300,000	Colombo	D 7	34
* Sudan	967,495	15,312,000	Khartoum	N 8	46
Surinam	55,144	389,000	Paramaribo	K 2	16
* Swaziland	6,704	411,879	Mbabane	N17	47
* Sweden	173,665	7,978,000	Stockholm	......	23
Switzerland	15,941	6,230,000	Bern	J 6	18
* Syria	71,498	5,866,000	Damascus	C 3	32
* Tanzania	362,819	12,896,000	Dar es Salaam	O13	47
Tennessee, U.S.A.	42,244	3,924,164	Nashville	M 5	7
Texas, U.S.A.	267,339	11,196,730	Austin	J 7	6
* Thailand	198,456	35,448,000	Bangkok	D 4	40
* Togo	21,853	2,004,711	Lomé	G10	46
Tokelau Islands	3.9	2,000	Fakaofo	J 6	44
Tonga	270	83,000	Nuku'alofa	J 8	44
* Trinidad and Tobago	1,980	1,040,000	Port of Spain	G 5	12
Tristan da Cunha	40	269	Edinburgh	H 7	1
* Tunisia	63,378	5,027,000	Tunis	J 4	46
* Turkey	301,381	34,375,000	Ankara	B 2	32
Turks and Caicos Is.	166	6,000	Cockburn Town	D 2	12
* Uganda	92,674	9,764,000	Kampala	N11	47
* Ukrainian S.S.R., U.S.S.R.	232,046	47,136,000	Kiev	D 5	30
* Union of Soviet Socialist Republics	8,649,498	241,748,000	Moscow	......	28,30
* United (Union of) Arab Emirates	32,278	179,126	Abu Dhabi	F 5	32
* United Kingdom	94,214	55,534,000	London	......	20
* United States of America, land	3,554,609	203,184,772	Washington, D.C.	......	6,7
land and water	3,615,123				
* Upper Volta	105,841	5,330,000	Ouagadougou	F 9	46
* Uruguay	72,172	2,900,000	Montevideo	J11	17
Utah, U.S.A.	84,916	1,059,273	Salt Lake City	F 4	6
Vatican City	109 acres	1,000		F 6	24
* Venezuela	352,143	10,398,907	Caracas	G 1	16
Vermont, U.S.A.	9,609	444,732	Montpelier	P 3	7
Vietnam, North	61,293	21,340,000	Hanoi	E 2	40
Vietnam, South	66,263	16,543,434	Saigon	E 5	40
Virginia, U.S.A.	40,817	4,648,494	Richmond	O 5	7
Virgin Islands, British	59	10,484	Road Town	H 1	12
Virgin Islands, U.S.A.	133	63,200	Charlotte Amalie	G 1	12
Wake Island, U.S.A.	2.5	1,097		G 4	44
Wales (incl. Monmouthshire), U.K.	8,017	2,724,540	Cardiff	E 4	20
Washington, U.S.A.	68,192	3,409,169	Olympia	C 3	6
Western Samoa	1,133	139,810	Apia	J 7	44
West Virginia, U.S.A.	24,181	1,744,237	Charleston	N 5	7
* White Russian S.S.R. (Byelorussian S.S.R.), U.S.S.R.	80,154	9,003,000	Minsk	C 4	30
Wisconsin, U.S.A.	56,154	4,417,933	Madison	L 4	7
World	57,491,000	3,632,000,000		......	1
Wyoming, U.S.A.	97,914	332,416	Cheyenne	G 4	6
* Yemen Arab Republic	75,000	5,000,000	San'a	D 7	32
* Yemen, Peoples Democratic Rep. of	111,075	1,220,000	Aden	E 7	32
* Yugoslavia	98,766	20,586,000	Belgrade	E 3	26
Yukon Territory, Canada	207,076	17,000	Whitehorse	C 3	3
* Zaire	905,563	21,637,876	Kinshasa	L12	47
* Zambia	290,586	4,056,995	Lusaka	M15	47

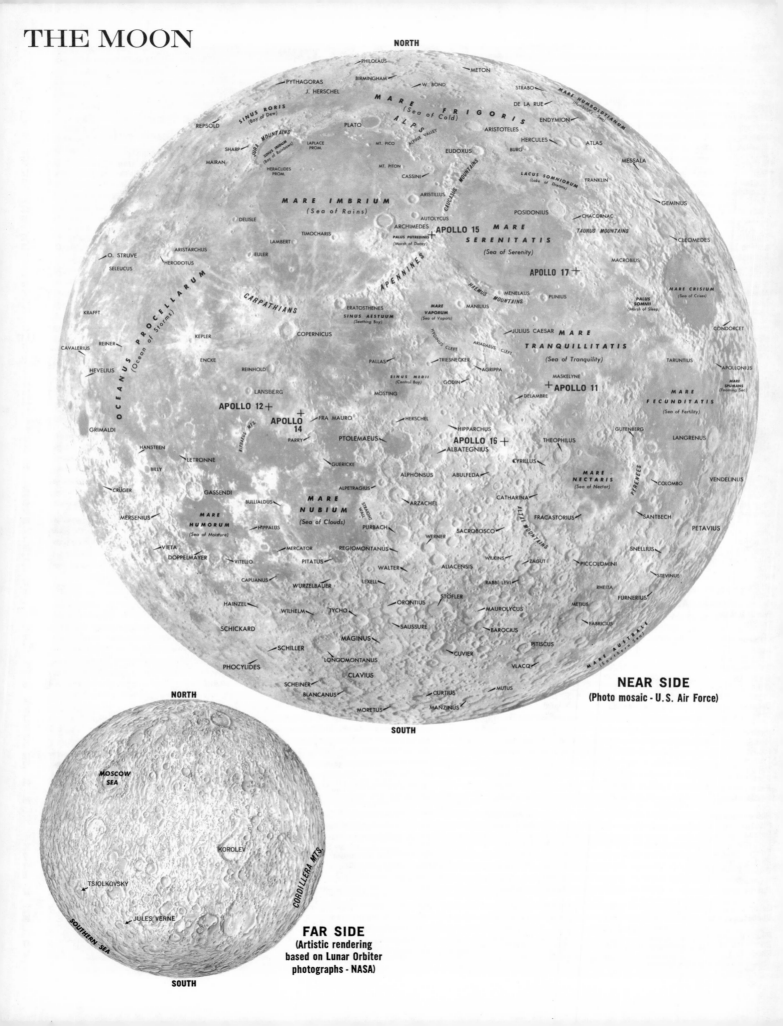

THE MOON

NEAR SIDE
(Photo mosaic - U.S. Air Force)

FAR SIDE
(Artistic rendering
based on Lunar Orbiter
photographs - NASA)